CLIMATE VARIABILITY – SOME ASPECTS, CHALLENGES AND PROSPECTS

Edited by **Abdel Hannachi**

Climate Variability – Some Aspects, Challenges and Prospects
Edited by Abdel Hannachi

Published by InTech
Janeza Trdine 9, 51000 Rijeka, Croatia

Publishing Process Manager Oliver Kurelic

Technical Editor Teodora Smiljanic

Cover Designer InTech Design Team

Additional hard copies can be obtained from orders@intechopen.com

Climate Variability – Some Aspects, Challenges and Prospects, Edited by Abdel Hannachi
p. cm.
ISBN 978-953-307-699-7

For EU product safety concerns:
IN TECH d.o.o., Prolaz Marije Krucifikse Kozulić 3, 51000 Rijeka, Croatia,
info@intechopen.com or visit our website at intechopen.com.

Meet the editor

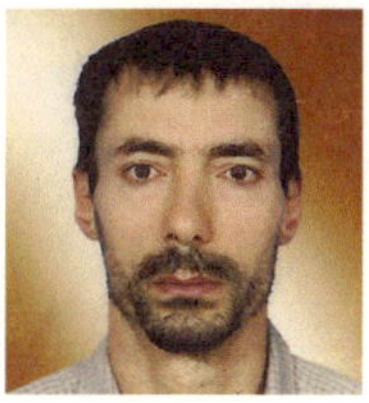

Abdelwaheb Hannachi is an associate professor in the Department of Meteorology at the University of Stockholm. Prior to the current position, he held various senior scientist positions, including those at Oxford and Reading Universities.

Contents

Preface

We are all familiar with the fascinating and ever changing weather. But as we grow older the memories we keep about weather is an accumulation of events. What we remember of weather after a long period of time, say few decades, is simply climate. Climate is in fact the statistics of weather, and as put by Ed Lorenz, climate, in mathematical terms, is the collection of all long-term statistical properties of the atmospheric state.

Climate varies on a wide range of scales both temporal and spatial, and we often talk about local, regional or global climate, but also decadal climate variability and long term trends. Climate is a complex high-dimensional system involving highly non-linear interactions between many processes. Climate variation is controlled by external factors such as the solar activity, eg the 11-year cycle, and cyclic variations in Earth's orbital parameters and also internal factors such as anthropogenic changes in greenhouse gases. Weather and climate variability (and change) has great impact on our society as well as the environment. There is a large wealth of scientific literature on climate variability and change and its impact on the infrastructure and society.

This book explores various perspectives about climate variability and change but is not meant to cover all aspects of the problem. The chapters in this book are divided into five sections. Section 1 consists of two chapters related to low frequency atmospheric variability. the first chapter provides a general but short review of the aspect of climate variabilitiy in two different climate regions: the North Atlantic/European sector in the extratropics and the Indian monsoon region in the tropics. Despite their climatic differences the two regions share common features related to nonlinearity where the atmospheric variability on intraseasonal time scales is characterised by an on-off switching between different weather regimes. Chapter 2 explores the land-atmosphere coupling strength using an ensemble of general circulation models from the Global Land-Atmosphere Coupling Experiment. Low-frequency atmospheric variability plays an important role in land-atmosphere coupling and precipitation predictability.

Section 2 consists of one chapter and explores the connection between solar activity and space weather and Earth's climate. Chapter 3 discusses both the supportive and controversial arguments of the solar effect on Earth's climate, and presents a different perspective to climate change.

Section 3 consists of two chapters related to one of the most important modes of climate variability: the El-Nino Southern Oscillation (ENSO). Chapter 4 discusses the benefits and prospects of ocean observation data asimilation for ENSO monitoring and seasonal forecasting. The quasi-coupled data assimilation system shows significant skill improvement of seasonal ENSO and atmospheric forecasting. Chapter 5 analyses the effect of one typical ENSO-type wind stress on the volume and heat transport variability in the world ocean. This type of wind stress can affect not only the mixed layer and the thermocline but also the thermohaline circulation.

Section 4 consists of two chapters and is devoted to rainfall, an important component of climate variability. Chapter 6 explores ways of assessing drought with particular application to Europe. The detrending method as well as the choice of the reference calibration period affect the sensitivity of drought assessment. Chapter 7 discusses a particular example of seasonal prediction of summer rainfall in the Bermejo river basin in Argentina. The southern annular mode phase and the Atlantic High have the most effect on summer rainfall in that region.

Section 5, with two chapters, is devoted to the adaptation to climate variability and change. Chapter 8 provides a review of adaptation strategies to face the risk of future climate change with a particular focus on the African continent. It summarizes the context of adaptation in a varying climate as opposed to a changing climate, and provides recommendations on the implementation strategy. Chapter 9 presents a succinct review of the process of technological change of innovation and its relation to a varying climate. It provides a new perspective to climate adaptation by taking into account environmental technology.

Abdel Hannachi
Department of Meteorology,
Stockholm University,
Sweden

Part 1

Atmospheric Variability

1

Atmospheric Low Frequency Variability: The Examples of the North Atlantic and the Indian Monsoon

Abdel Hannachi[1], Tim Woollings[2] and Andy Turner[3]
[1]*Department of Meteorology, Stockholm University, Sweden*
[2]*Department of Meteorology, University of Reading*
[3]*NCAS-Climate, Walker Institute for Climate System Research, Department of Meteorology, University of Reading*
[1]*Sweden*
[2,3]*UK*

1. Introduction

Great efforts, sometimes taking the form of a race, are exerted by climate scientists to provide medium and long-term future climate predictions on large and regional, or even local scales. This exercise has proved to be a really challenging one. There is a wide variety of climate characteristics between different regions on the globe. For example, tropical and subtropical regions tend to be more influenced by what happens in the equatorial Indian and Pacific oceans such as the El Nino Southern Oscillation (ENSO). Midlatitude regions, on the other hand, are more affected by the north-south migration of the polar front or synonymously the midlatitude jet stream. It is important to notice that even within the midlatitude band climate variation differs from region to region. For example, climate variability over the North Atlantic European region is different from that of the Pacific North America (PNA) region and is particularly more difficult to predict.

The jet stream is a belt of strong westerly wind that goes around the globe in the subtropics (subtropical jet) or the midlatiudes (eddy-driven jet). The subtropical jet results from the westerly acceleration of poleward moving air associated with the upper branch of the Hadley cell. The midlatitude jet stream, on the other hand, results from the momentum and heat forcing by midlatitude eddies, i.e. weather systems. Weather and climate variations in the extratropics are associated to a large extent with meridional shifts of the midlatitude westerly jet stream. For instance, major extratropical teleconnections, including the North Atlantic Oscillation (Fig. 1) and the PNA pattern, describe changes in the jet stream (Wittman et al. 2005; Monahan and Fyfe 2006). Over the North Atlantic region, the North Atlantic Oscillation (NAO) is the dominant large scale mode of variability with its north-south dipole anomaly centres (Hurrell et al. 2002). It is a seesaw in atmospheric mass between the subtropical high and and the polar low and affects much of the weather and climate in the North Atlantic, east of North America, Europe and parts of Russia. The positive phase of the NAO (Fig. 1b) is generally associated with a stronger subtropical high pressure and a deeper than normal

Icelandic low yielding warmer and wetter, than normal, conditions over Europe associated with colder and drier, than normal, conditions in northern Canada and Greenland. The negative phase (Fig. 1a) is the opposite of the positive phase and yields moist air into the Mediterranean and cold air in northern Europe.

The second prominent mode of variability over the North Atlantic-European region is the East Atlantic (EA) pattern. The EA pattern also has a north-south dipole of anomaly centres that are displaced southward with respect to those of the NAO so that both patterns are in quadrature and the northern centre, centered around 45^oN, is stronger than the lower latitude centre, which is more linked to the subtropics and modulated by the subtropical ridge. The positive phase of the EA is associated with above- and below-average surface temperature over Europe and eastern North America respectively. The variability of these modes is usually described by patterns in pressure or geopotential height fields, or wind fields as in Athanasiadis et al. (2009). Jet stream shifts are associated with a positive feedback between the mean flow and the transient eddies (eg, Lorenz and Hartmann 2003).

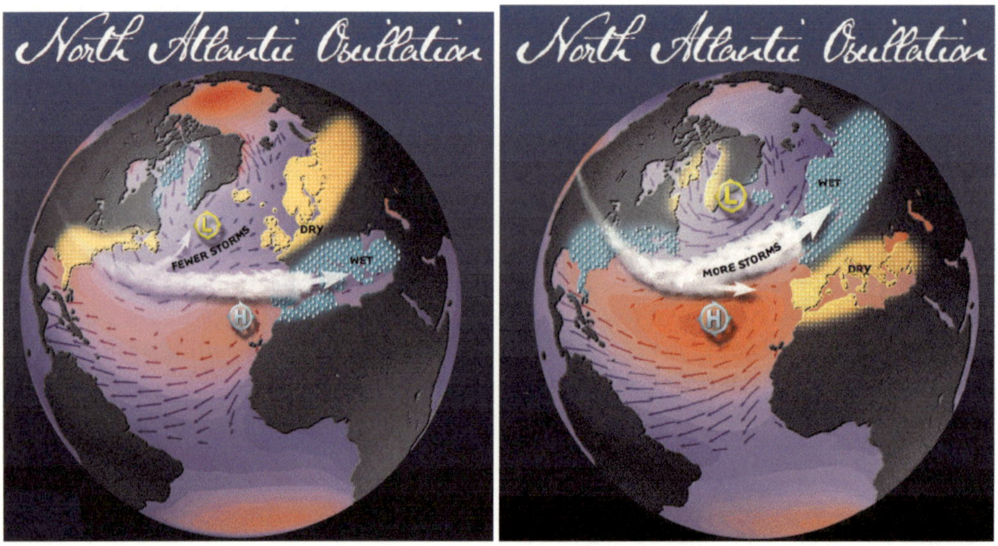

(a) Negative NAO (b) Positive NAO

Fig. 1. Illustration of the negative (a) and positive (b) phases of the NAO pattern in terms of winds, moisture and surface temperature. Source: http://www.ldeo.columbia.edu/res/pi/NAO/.

Woollings et al. (2010a, WO10a hereafter) analysed the variability of the leading mode of the 500-hPa geopotential height (Z500) derived from the 44 winters (December-February, DJF) 1957/58-2000/01 of the 40-year European Centre for Medium-Range Weather Forecasts (ECMWF) Re-Analysis (ERA-40) (Uppala et al. 2005). They suggested that the NAO can be interpreted in terms of a transition between two states; a high-latitude (Greenland) blocking and a no blocking flow. The complex behaviour of the jet stream variability means that it requires at least two spatial patterns to describe its dominant variations (Fyfe and Lorenz 2005; Monahan an Fyfe 2006), and for the North Atlantic these are the NAO and the EA patterns (Woollings et al. 2010b). Woollings et al. (2010b, WO10b hereafter) considered the winter (DJF) ERA-40 low-level (925-700 hPa) wind to analyse the latitude and speed of the eddy-driven jet stream. Their analysis suggests, as it is also described below, that there are three preferred latitudinal positions of the North Atlantic jet stream, and this is in very good agreement with similar flow structures obtained from a Gaussin mixture model applied to the two-dimensional (NAO,EA) state space. Two of the jet positions are found to be associated with the two states identified in WO10a, and which reflect the NAO variability.

Climate is by definition a high dimensional and complex system involving highly nonlinear interacting processes. Nonlinearity means, in particular, that changes in climate due to changes in external forcing, such as greenhouse gases, do not scale linearly with the latter and surprises are expected to occur. Weather and climate variability is not pure randomness but embeds some sort of dynamical structure. In synoptic meteorology, for example, it has been the practice to regard weather and climate variability as consisting of a small number of large scale weather patterns, also known as weather regimes, that reccur intermittently hence affecting regional climate through their persistence and integrating effect. Persistence and meridional shifts of the jet stream could therefore hold the key to any regime-like behaviour. Under climate warming these regimes are expected to change by changing their structure and/or their frequency of occurrence (Palmer Palmer 1999; Branstator and Selten 2009). These changes can have serious impacts on the economy and society. For example, under global warming it is projected that deserts and areas susceptible to drought will increase. In the meantime extreme precipitation events, which often damage crops, and (summer) heat waves, which cause health problems, will become more frequent.

In the tropics different processes are involved. For the monsoons, for example, the fundamental driving mechanisms are differential heating between sea and land masses and moisture transport. One of the main regions of monsoon activity on Earth is the Asian monsoon region. The Asian summer monsoon is very important not least for affecting the lives of more than the third of the world's population. While seasonal mean Asian monsoon is reasonably well understood through lower-boundary forcings (Charney and Shukla 1981), subseasonal (30-60 day timescale) variations of monsoon or monsoon intraseasonal variability (MISV), generally linked to what is commonly known as active and break monsoon phases, is less so. Although MISV tends to be more chaotic there is evidence suggesting increased frequency of active (break) conditions during strong monsoon (drought) years.

This chapter reviews and discusses the state-of-the-art of climate variability and nonlinearity in the midlatitude and the tropical regions based on the works of Woollings et al. (2010b) and Turner and Hannachi (2010, TH10). We show, in particular, the similarirty between the two regions in terms of nonlinearity and the possible effect of global warming using ERA-40 reanalyses. The first region is the winter North Atlantic European sector characterised by its midlatitude climate (WO10b). The second one is the summer monsoon region around India and South East Asia (TH10). Both regions are found to be characterised by nonlinear regime behaviour. The study applies mixture model techniques (Hannachi and Turner 2008; TH10; WO10b) to the jet latitude index and the NAO/EA teleconnection patterns in one case and to a simple index of the Asian summer monsoon convection derived from the ERA-40 reanalysis in the other. Section 2 describes the data and methodology. Section 3 discusses the case of the North Atlantic/European region and section 4 discusses the Asian summer monsoon case. A summary and a discussion are presented in the last section.

2. Data and methodology

2.1 Data

We have used the 500-hPa geopotential height (Z500) data from the ERA-40 reanalysis project (Uppala et al. 2005). The gridded data are defined on a regular $2.25^o \times 2.25^o$ grid north of 20^oN and span the period December–February (DJF) 1957/58–2000/01 yielding 44 complete winter (DJF) seasons. Daily and monthly data are used. A smooth seasonal cycle is obtained by averaging daily data over all the years then smoothing with a discrete cosine transform retaining only the mean and the lowest two Fourier frequencies. Daily anomalies are obtained by subtracting the smooth seasonal cycle from the original daily data.
For the Asian monsoon, we have used the outgoing long-wave radiation (OLR) and 850-mb wind fields from ERA-40 (Uppala et al. 2005) over the Asian summer monsoon region $(50 - 145E, 20S - 35N)$ for the period 1958-2001. Daily detrended anomalies are obtained by removing the seasonally-varying mean field based on monthly averages. The monsoon season is defined by the months June-September (JJAS). In addition, to characterise the large scale seasonal mean influence on monsoon convection we have used the dynamical monsoon index (WY) proposed by Webster and Yang (1992). The WY index is a proxy for the (adiabatic) heating of the atmospheric column and is defined as the JJAS average of anomalous zonal wind shear between the lower (850-hPa) and upper (200 hPa) tropospheric levels averaged over the band $(40 - 110E, 5 - 20N)$. We also used the daily India Meteorological Department rainfall gridded data (Rajeevan et al. 2006) as an independent measure of monsoon rainfall (see TH10 for more details).

2.2 Methodology

The jet-latitude index (WO10b) is computed for the period 1 December 1957 – 28 February 2002 by averaging daily mean zonal winds over the levels 925, 850, 775 and 700 hPa and the longitudes $0 - 60^o$W. A 10-day low-pass Lanczos filter is then applied to the data and the maximum wind speed value is used to define the jet latitude and speed. A smooth seasonal cycle is then removed from these to give anomaly values (see WO10b for more details). The NAO and EA patterns and associated indices are obtained as the leading empirical orthogonal functions (Hannachi et al. 2007) of Z500 anomalies over the Atlantic sector $(20^o - 90^o N, 90^o W - 90^o E)$, see WO10a and WO10b for details.
To estimate the probability density function (PDF) function we used the unidimensional kernel density estimation method (Silverman 1981). In addition we have also used the univariate and multivariate mixture model approach (Hannachi 2007; WO10b; TH10). In this model, the PDF is decomposed as a weighted sum of Gaussian (univariate and multivariate) normal PDFs. The centre and the covariance structure of each Gaussian component from the mixture is then analysed separately.

3. North Atlantic jet and atmospheric circulation

As we have mentioned in the introduction, the NAO is the dominant mode of weather and climate variability over the North Atlantic sector. WO10a showed that the NAO can be explained as a transition between two flow states (Fig. 2); a Greenland blocking (GB), associated with a negative NAO phase, and a no-blocking flow, looking like a split jet and is associated with a positive phase of the NAO. Croci-Maspoli (2007) showed, in fact, that when all blocking events were removed from the ERA40 the NAO is no longer the leading empirical

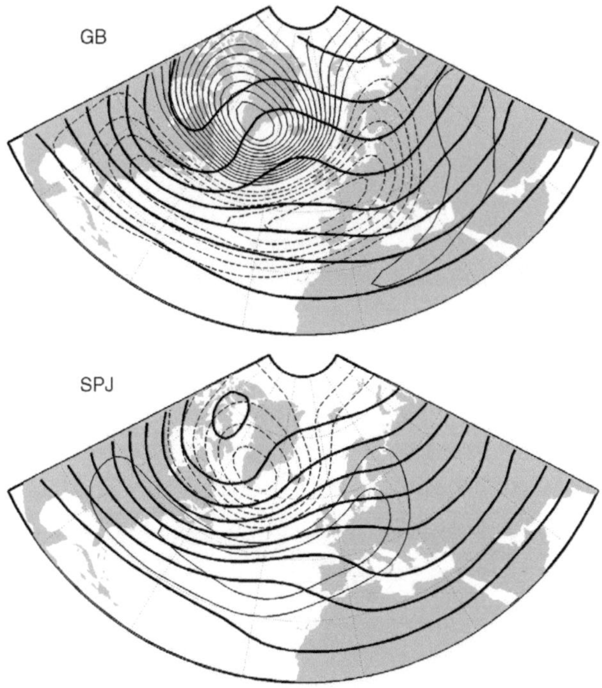

Fig. 2. Flow regimes of the full (thick contours) and anomaly (thin contours) of the winter (DJF) ERA-40 Z500 field obtained from the mixture model applied to the NAO time series. Contour intervals: 100 m (full field) and 10 m (anomalies). Negative contour dashed (reproduced from WO10a)

orthogonal function or EOF (Hannachi et al. 2007). Since much of the weather and climate variability of the extratropics is associated with the jet stream variability it is expected that the flow states or regimes must be associated somehow with particular structures of the jet. WO10b identified the eddy-driven jet stream by computing the jet latitude and the associated maximum wind.

Figure 3 shows an example of the time evolution of the jet latitude for the four winters (DJF) 1957/1958 to 1960/1961 of the zonal wind computed over the North Atlantic region. The jet latitude is characterised by periods of persistence at specific latitudes[1] and periods of transitions between these latitudes. This indicates that the jet stream is characterised by persistence in addition to north-south migration. An extended period of the jet evolution over 8 winters Dec 1958 - Feb 1967 is shown in Fig. 4a as a time series. To identify the persistence locations of the eddy-driven jet stream Fig. 4b shows the kernel PDF of the jet latitude along with the same PDF estimated using the mixture model. The jet latitude PDF clearly has a trimodal structure reflecting three preferred locations for the North Atlantic eddy-driven jet stream shown by the dotted lines in Fig. 3.

The Z500 anomaly flow patterns associated with the peaks of the jet latitude PDF are shown in Fig. 5 based on compositing over the closest 300 days to these peaks. The left hand side peak corresponds to the southern jet regime characterised by its high pressure or blocking over

[1] shown by the dotted lines in Fig. 3 and are discussed later

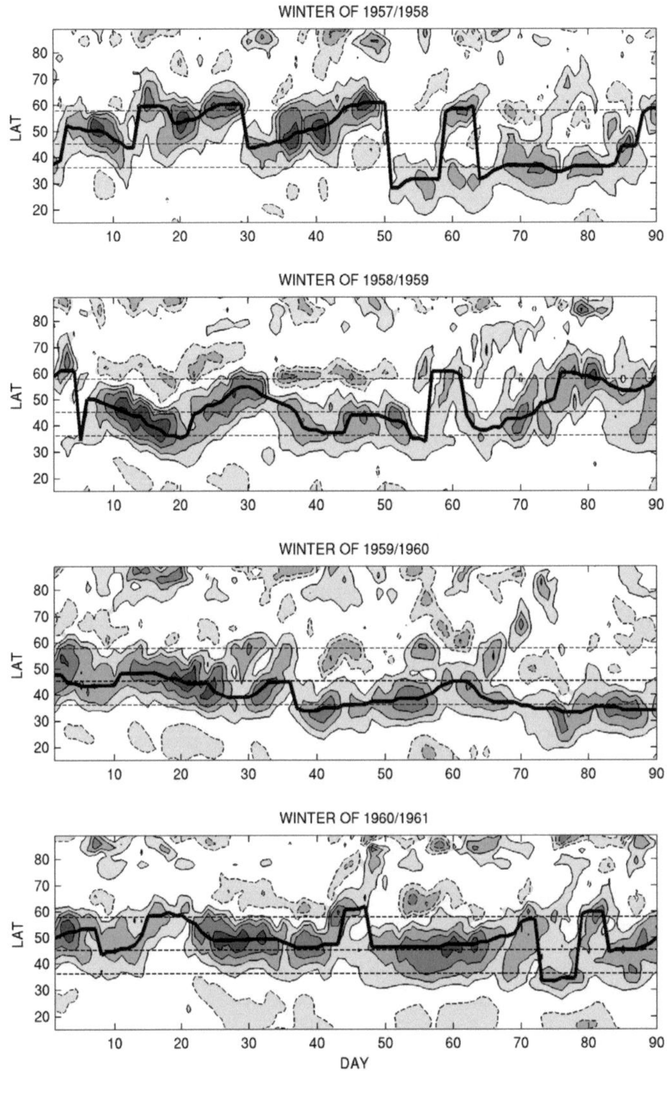

Fig. 3. Daily zonal-mean zonal-wind averaged over longitudes $0 - 60^{o}$W and pressure levels 925-700 hPa versus time for the first four years of ERA-40 winters (DJF 1957/1961). The jet latitude is shown by the thick line and the preferred jet latitudes are shown by dashed horizontal lines. Contour interval 5 m/s, and negative contours dashed.

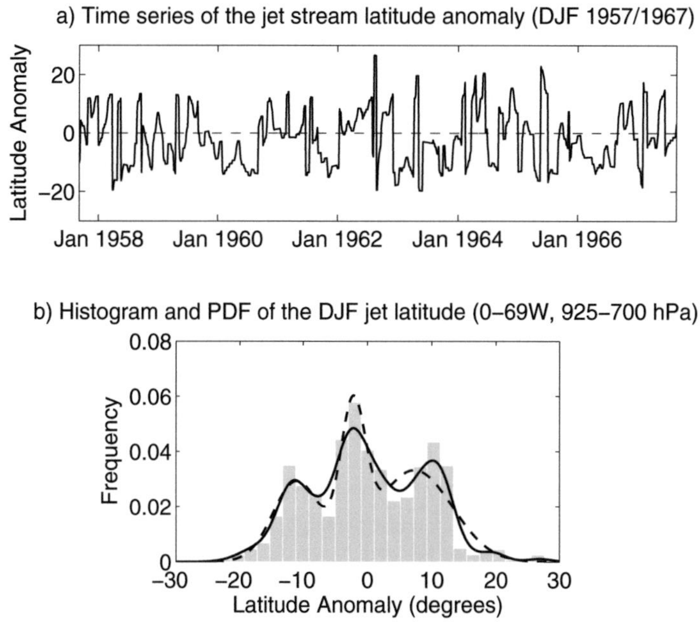

Fig. 4. A segment of the jet latitude time series for the first 10 winters (DJF 1957/1967) of ERA-40 (a), and the histogram along with the kernel (continuous) and mixture model (dashed) PDF estimate (b). Reproduced from Hannachi et al. (2012).

Greenland. The middle peak is associated with a low pressure centre over the midlatitude North Atlantic whereas the right hand side peak corresponds to a high pressure over the midlatitude North Atlantic. The southern jet position is similar to the negative NAO phase whereas the middle and north jet regimes look more like the opposite phases of the EA pattern. A similar composite applied to the zonal wind (not shown) indicates that only for the central and north jet composites is the eddy-driven jet stream separated from the subtropical jet (WO10b).

To link the jet variation to the low frequency variability in the North Atlantic/European sector we consider next the state space of the winter (DJF) daily Z500 anomalies, and we follow WO10b by using the leading two modes of variability, i.e. the NAO and the EA patterns. Fig. 6 shows a scatter plot of the data color-coded to show the latitude (anomaly) of each day. The mixture model is applied as in WO10b to this scatter plot using three bivariate Gaussian components each characterised by its centre (or mean) and its covariance matrix.

The ellipses in Fig. 6 reflect the covariance structure of the different bivariate components and the small filled circles represent their centres. The projection onto the NAO-EA plane of the patterns shown in Fig. 5 are indicated by crosses and they are quite close to the centres of the mixture components. In addition the ellipses are also in very good agreement with the colors of the data points (Fig. 6). The Z500 anomaly maps of the centres of the mixture model are shown in Fig. 7. These regimes are very similar to the composites of the jet regimes shown in Fig. 5.

It is clear that the low-frequency flow regimes over the North Atlantic sector are associated to the persistent states of the eddy-driven jet stream. The southern jet position is explained by the persistent GB blocking. The central position seems to be related to the undisturbed state

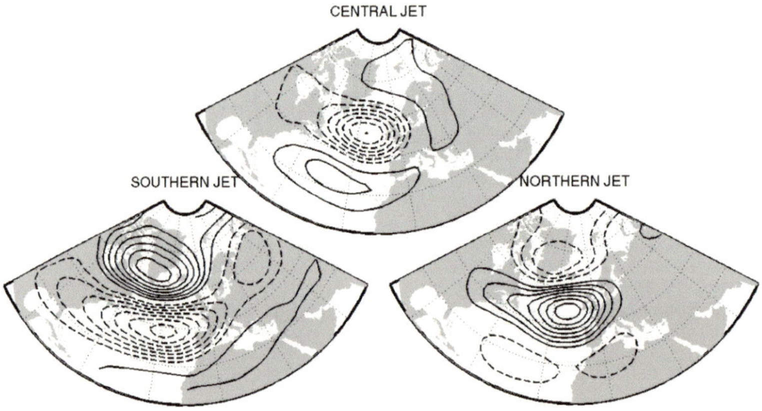

Fig. 5. 500mb geopotential height maps corresponding to the three PDF peaks of the jet latitude. Contour interval 20 m, negative contours dahed and zero contour omitted. Reproduced from WO10b.

of the jet given its proximity to the mode (or peak) of the two dimensional Gaussian mixture distribution (not shown). As for the northern jet position, it only partly reflects the occurrence of blocking in the southwest of Europe, which could divert the jet northward (Woollings et al. 2011).

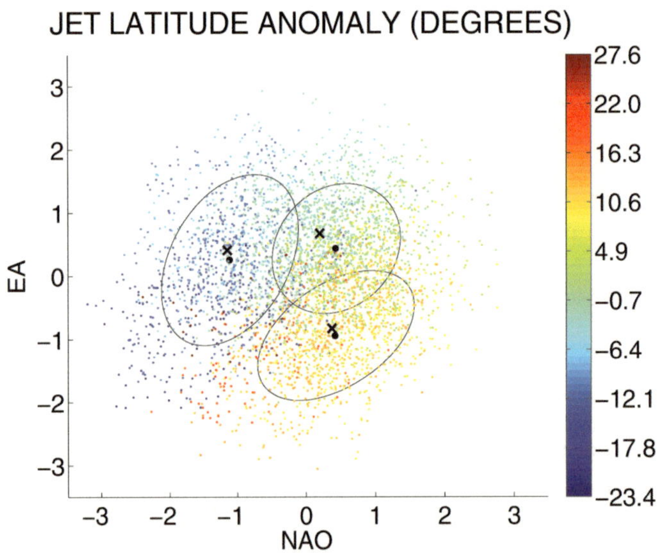

Fig. 6. Scatter plot of the daily winter (DJF) Z500 anomalies, projected onto the NAO/EA plane and color-coded to show the jet latitude. The ellipses and their associated centres correspond respectively to the covariances and the means of the Gaussian component mixture model. The crosses represent the regimes obtained from the jet latitude PDF projected onto the same plane (reproduced from WO10b).

An important issue arises in climate variability in relation to global warming, and that is the following. How will weather and climate variability look like in a warmer future? This is an important question for strategic planning. The available reanalyses data are generally limited to less than 100 years, including the ERA-40, which is only about 50 years long, and therefore cannot be used to give a definite answer to the above question. We have, never the less, attempted to look at this question by splitting the jet latitude time series into pre- and post-1978 subsamples and looked at the respective PDFs. The result (not shown) indicates that the trimodal structure is conserved between the two periods. There is, however, a significant decrease of the southern jet regime frequency in the last half of the record compared to the first half. This is concomitant with a decrease of Greenland blocking frequency. We reiterate again that, given the length of the data, this could simply reflect the natural variability rather than an anthropogenic trend.

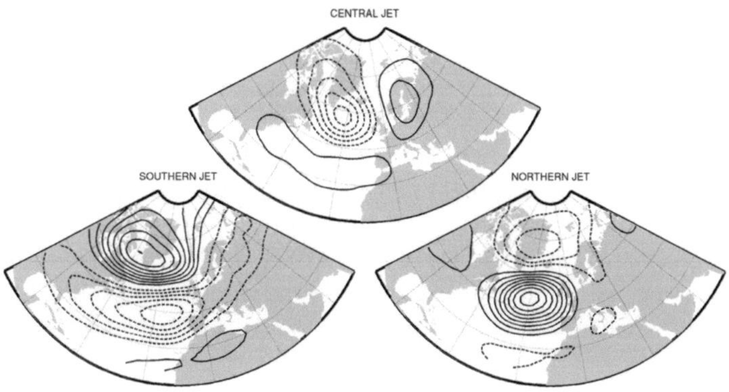

Fig. 7. As in Fig. 5 but for the centres of the Gaussian components of the mixture model. Reproduced from WO10b.

There is also a slight increase of frequency of the central and northern jet frequency. As for the latitudinal shift there is a slight hint of a northward shift of the jet latitude PDF peaks although it is not significant. Climate change studies based on the climate model intercomparison project (CMIP3) models (Barnes and Hartmann 2010) do indicate indeed a northward shift of the jet stream in warmer climate. Despite the rather large differences between the climate models of the CMIP experiment the northward shift of the jet stream seems to be a robust feature.

4. Asian monsoon variability

The OLR is a proxy for convection and we use it here to discuss the MISV over the Asian summer monsoon region. The leading EOF of the OLR anomalies explain about 24% of the total variance and is well separated from the variances of the rest of the modes of variability and we discuss the MISV based on this mode of variability following TH10. Figure 8a shows the OLR EOF1 with its dipolar structure showing opposite variability between the maritime continent and parts of India and south China. The first principal component (PC1) associated with EOF1 (Fig. 8a) is used to analyse MISV. Fig. 8b shows the PDF of the index along with the two Gaussian components used in the two-component mixture model.
The left hand side regime R1 (Fig. 8b) is clearly associated with the opposite phase of the EOF1 pattern (Fig. 8a), i.e. a negative phase of OLR over southern India associated with a positive phase over the maritime continent. The composites of daily 850-mb wind and OLR

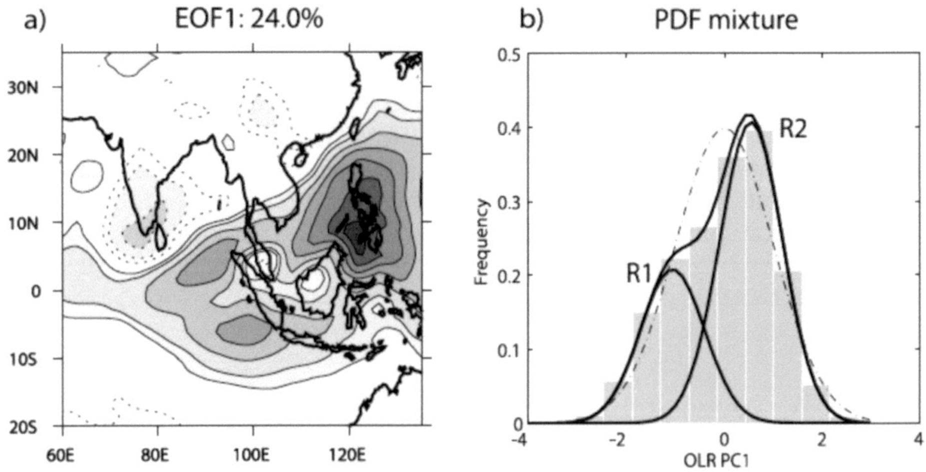

Fig. 8. (a) Leading empirical orthogonal function of JJAS ERA-40 OLR anomalies for the period 1958-2001. (b) Probability density function (upper solid curve) of the OLR index and the associated two Gaussian components of the mixture model (lower solid curves, indicated by R1 and R2). The dashed-dotted curve represents the Gaussian PDF fitted to the index. In (a) the contour interval is arbitrary and positive (negative) contours are dotted (continuous) (reproduced from TH10).

anomaly fields based on days close to the PDF mode corresponding to the left regime R1 (Fig. 8b) are shown in Fig. 9a. A similar composite of rainfall for the same regime R1 is shown in Fig. 9b. The regime flow R1 is consistent with an anticyclonic circulation over the maritime continent and south China sea with a reversal of the Somali jet and a diversion northward with convergence over most of the southern part of India. The composite map of rainfall (Fig. 9b) clearly shows a positive precipitation anomaly consistent with the monsoon active phase. The second regime R2 (Fig. 8b) has an opposite OLR phase to that of R1 with a positive OLR phase over southern India and a negative phase over the maritime continent. The wind field composite (Fig. 9c) shows a divergent flow over india and eastern Bay of Bengal and a cyclonic circulation over the Philippines and South China sea. The OLR amplitudes (Fig. 9c) are smaller than those of R1, with about 5 w/m^2 vs 15 w/m^2 over southern India and the Philippines respectively. The wind field is also weaker with a southward shift of the Somali jet. The map of rainfall composites (Fig. 9d) shows dry conditions over India consistent with a break phase of the Summer Asian monsoon. The robustness of the active and break phases has been tested in TH10, to which the reader is referred for more details.

The trend analysis of MISV was investigated by comparing monsoon activity between the first and second halves of ERA-40 data (TH10). The results indicate that the active monsoon has been reduced whereas the break phase has become more frequent in the second half (Fig. 10a). The relationship between the intraseasonal monsoon and the large-scale seasonal mean monsoon was also addressed by TH10 using the Webster-Yang (WY) index. We found that MISV is closely related to the large-scale monsoon variability (Fig. 10b). Precisely, seasons with above-normal monsoon heating the break and active phases have equal likelihood. On the other hand, seasons with below-normal large-scale monsoon heating the break phase becomes more likely.

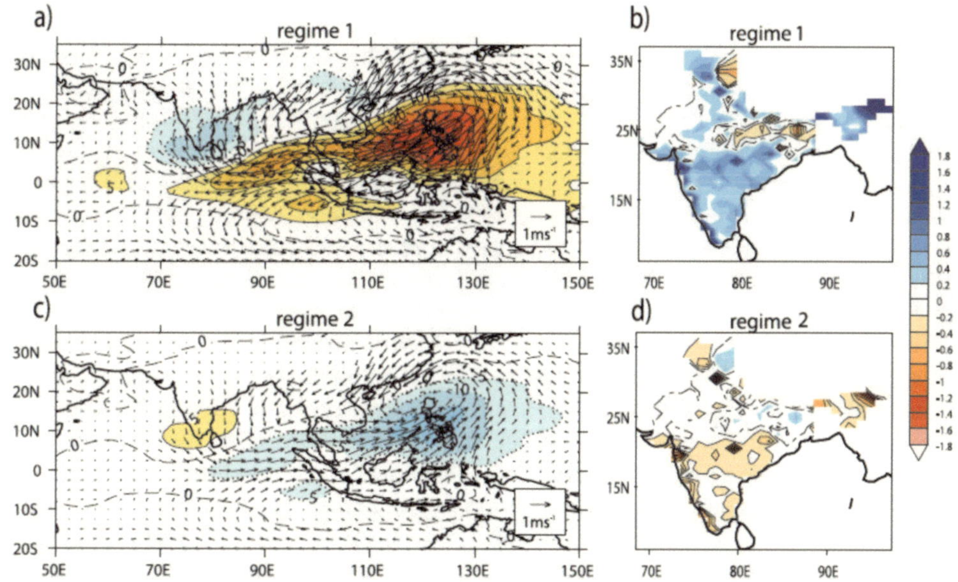

Fig. 9. Composite anomalies of OLR and wind field and rainfall over India for the first (a,b) and second (c,d) monsoon regimes over the 1958-2001 period. Contour interval for OLR composites is 5 wm^{-2}, red solid (blue dotted) is positive (negative). Rainfall contours are 0.2 mm/day, and negative contour lines only are shown (reproduced from TH10).

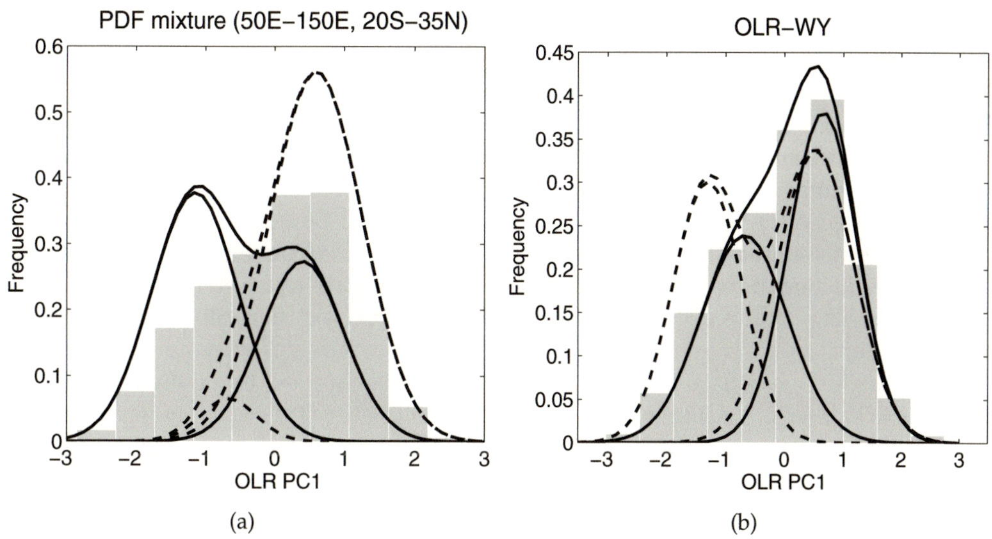

Fig. 10. (a) PDFs of the daily OLR time series in the early (solid) and late (dahsed) part of the record. Upper (lower) curves represent total (mixture components) probability distributions. The left (right) component of the mixture represents regime R1 (R2). (b) Perturbations to the whole period mixture when the OLR index is stratified by JJAS-average dynamical monsoon, or Webster-Yang, index: WY^{+} (dashed) and WY^{-} (solid). Reproduced from TH10.

5. Summary and conclusion

We have reviewed in this chapter some specific characteristics relating to the nature of the large scale and low-frequency atmospheric variability. The discussion focussed essentially on the nonlinear nature governing this variability. Two important regions are discussed in this chapter, one in the midlatitudes and the other in the tropics. The first region is the North Atlantic/European sector and the discussion follows Woollings et al. (2010b). The second region is the summer monsoon Asia, and the discussion follows Turner and Hannachi (2010). The data used come from the ECMWF and consist of the ERA-40 winter (DJF) geopotential height, zonal wind from the lower troposphere, over the North Atlantic sector as well as the 850 mb horizontal wind and sea level pressure over the Asian summer (JJAS) monsoon region. In the extratropics well-documented prominent modes of variability are known to control the climate variability. The North Atlantic Oscillation, a north-south seeesaw in the atmospheric mass, and also the East Atlantic pattern constitute major contributors to weather and climate variation in the North Atlantic/European region. For the NAO pattern, for example, various mechanisms have been proposed to explain its existence. WO10a, for example, suggested that the NAO is the consequence of regime transitions between the two flow patterns; a Greenland blocking and a no-blocking flow. Weather and climate variability in the extratropics can be explained by variation, such as meriodinal shift, of the midlatitude westerly jet stream. It is therefore natural to seek an explanation of the low-frequency and large-scale flow patterns in the North Atlantic sector based on the midlatitude or eddy-driven jet stream variability.
A jet latitude index was compued by WO10b, based on the maximum of the zonal mean zonal wind averaged over the North Atlantic sector and the four lowest pressure levels of ERA-40. The PDF of the jet latitude was then computed and revealed a trimodal structure. The modes represent three latitudinal positions of the eddy-driven jet stream. The first one represents the southern jet position, situated around latitude 36^oN, and is associated with the Greenland blocking. The next one represents the middle jet position around 45^oN, and the last one represents the jet when it is at its northern most latitude, around 57^oN.
These jet locations have been linked to the weather and climate variability over the sector. Using the reduced state space spanned by the two leading modes of variability, the NAO and EA patterns, of the 500-mb geopotential height, the mixture model yields three regimes very similar to those associated with the peaks of the jet latitude PDF. A simple analysis based on comparing the jet latitude time series between the two halves of the ERA-40 record reveals a significant reduction of the frequency of the southern jet regime as we go from the first to the second half of the data record. In addition, there is a northward shift, albeit small, of the jet stream location.
The same analysis, based on the mixture model, was also applied to the time series of the first OLR EOF over the Asian monsoon region. Two phases of the intraseasonal monsoon variability were identified, which are consistent with the break and active monsoon phases over India. The seasonal mean condition is then found to affect the likelihood of these regimes providing evidence that large scale forcing can lend some predictability to monsoon weather patterns during the season. For example, seasons with above-normal monsoon heating can yield equal likelihood for both intraseasonal monsoon phases. The trend analysis of intraseasonal monsoon activity also reveals an increase of the break phase at the expense of the active phase. A more detailed analysis of these issues is, however, beyond the scope of this chapter and is left for future research.

6. Acknowledgements

We thank ECMWF for providing the ERA-40 reanalysis data.

7. References

Athanasiadis, P. J.; Wallace, J. M., & J. J. Wettstein, J. J. (2009). Patterns of jet stream wintertime variability and their relationship to the storm tracks. *Journal of the Atmospheric Sciences*, Vol., 67, 1361–1381.

Barnes, E. A. & Hartmann, D. L. (2010). Influence of eddy-driven jet latitude on North Atlantic jet persistence and blocking frequency in CMIP3 integrations. *Geophysical Research Letters*, Vol., 37, doi:10.1029/2010GL045700.

Branstator, G. & Selten, F. (2009). "Modes of Variability" and Climate Change. *Journal of Climate*, Vol., 22, 2639–2658.

Charney, J. G. & Shukla, J. (1981). Predictability of monsoons, in Sir Lighthill, J.; & Pearce, R. P. (ed.), *Monsoon Dynamics*, Cambridge University Press, pp. 99–109.

Croci-Maspoli, M.; Schwierz, C. & Davies, H. C. (2007). Atmospheric blocking: Space-time links to the NAO and PNA. *Climate Dynamics*, Vol, 29, 713–725.

Fyfe, J. C. & Lorenz, D. J. (2006). Characterizing midlatitude jet variability: lessons from a simple GCM. *Journal of Climate*, Vol, 18, 3400–3404.

Hannachi, A.; Jolliffe, I. T. & Stephenson, D. B. (2007). Empirical orthogonal functions and related techniques in atmospheric science: A review. *International Journal of Climatology*, Vol, 27, 1119–1152.

Hannachi, A. (2007). Tropospheric planetary wave dynamics and mixture modeling: Two preferred regimes and a regime shift. *Journal of the Atmospheric Sciences*, Vol, 64, 3521–3541.

Hannachi, A. & Turner, A. G. (2008). Preferred structures in large scale circulation and the effect of doubling greenhouse gas concentration in HadCM3. *Quarterly Journal of the Royal Meteorological Society*, Vol, 134, 469–480.

Hannachi, A.; Woollings, T. & Fraedrich, K. (2012). The North Atlantic jet stream: A look at preferred positions, paths and transitions. *Quarterly Journal of the Royal Meteorological Society*, in press.

Hurrell, J. W.; Kushnir, Y.; Ottersen, G. & Visbeck, M. (2002). An overview of the North Atlantic Oscillation. In *The North Atlantic Oscillation—Climate Significance and Environmental Impact,* Geophysical Monograph, Vol, 134, American Geophysical Union, 1–35.

Lorenz, D. J. & Hartmann, D. L. (2003). Eddy-Zonal Flow Feedback in the Northern Hemisphere Winter. *Journal of Climate*, Vol, 16, 1212–1227.

Monahan, A. H. & Fyfe, J. C. (2006). On the nature of zonal jet EOFs. *Journal of Climate*, Vol, 19, 6409–6424.

Palmer, T. (1999). A nonlinear dynamical perspective on climate prediction. *Journal of Climate*, Vol, 12, 575–591.

Rajeevan, M.; Bhate, J., Kale, J. D. & Lal, B. (2006). High resolution daily gridded rainfall data for the Indian region: Analysis of break and active monsoon spells. *Current Science*, Vol, 91, 296–306.

Silverman, B. W. (1981). Using kernel density estimates to investigate multimodality. *Journal of the Royal Statistical Society*, Vol, 43, 97–99.

Turner, A. G. & Hannachi, A. (2010). Is there regime behavior in monsoon convection in the late 20th century? *Geophysical Research Letters*, Vol, 37, doi:10.1029/2010GL044159.

Uppala, S. M. & Coauthors, (2005). The ERA-40 Re-Analysis. *Quarterly Journal of the Royal Meteorological Society,* Vol, 131, 2961–3012.

Webster, P. J. & Yang, S. (1992). Monsoon and ENSO–Selectivity interactive systems. *Quarterly Journal of the Royal Meteorological Society,* Vol, 118, 877–926.

Wittman, M. A. H.; Charlton, A. J. & Polvani, L. M. (2005). On the meridional structure of annular modes. *Journal of Climate*, Vol, 18, 2119–2122.

Woollings T.; Hannachi, A.; Hoskins, B. J. & Turner, A. G. (2010a). A regime view of the North Atlantic Oscillation and its response to anthropogenic forcing. *Journal of Climate,* Vol, 23, 1291–1307.

Woollings T.; Hannachi, A. & Hoskins, B. J. (2010b). Variability of the North Atlantic eddy-driven jet stream. *Quarterly Journal of the Royal Meteorological Society,* Vol, 136, 856–868.

Woollings, T., Pinto, J. G. & Santos J. A. (2011) Dynamical Evolution of North Atlantic Ridges and Poleward Jet Stream Displacements. *J. Atmos. Sci.*, 68, pp. 954-963.

2

Impact of Atmospheric Variability on Soil Moisture-Precipitation Coupling

Jiangfeng Wei[1], Paul A. Dirmeyer[1], Zhichang Guo[1] and Li Zhang[2]
[1]*Center for Ocean-Land-Atmosphere Studies,*
Institute of Global Environment and Society,
Calverton, Maryland,
[2]*NOAA/NWS/NCEP/Climate Prediction Center,*
Camp Springs, Maryland
USA

1. Introduction

It is now well-established that the chaotic nature of the atmosphere severely limits the predictability of weather, while the slowly varying sea surface temperature (SST) and land surface states can enhance the predictability of atmospheric variations through surface-atmosphere interactions or by providing a boundary condition (e.g., Shukla 1993, 1998; Shukla et al. 2000; Graham et al. 1994; Koster et al. 2000; Dirmeyer et al. 2003; Quan et al. 2004). Among them, the influence of ocean is more important on a global scale because it covers twice as much surface area as land and is a much larger heat and energy reservoir. But the impact of ocean may not be dominant over land, especially the mid-latitude land (Koster and Suarez 1995).

The Global Land-Atmosphere Coupling Experiment (GLACE) (Koster et al., 2004, 2006) builds a framework to objectively estimate the potential contribution of land states to atmospheric predictability (called land-atmosphere coupling strength) in numerical weather and climate models. By averaging the estimated land-atmosphere coupling strength from 12 models participating in GLACE, an ensemble average coupling strength is obtained. However, the coupling strength varies widely among models. The discrepancy is certainly related to differences in the parameterization of processes and their complex interactions, from soil hydrology, vegetation physiology, to boundary layer, cloud and precipitation processes. It is difficult to determine what causes the relatively strong or weak coupling strengths seen in individual models.

Some studies have identified the impact of soil moisture on evapotranspiration (ET) (denoted SM→ET) and the impact of ET on precipitation (denoted ET→P) as two key factors for land-atmosphere coupling (Guo et al. 2006 (hereafter GUO06); Dirmeyer et al. 2010). For soil moisture to have a strong impact on precipitation, both SM→ET and ET→P need to be strong. This usually happens in transitional zones between wet and dry climates (Dirmeyer 2006). In addition to the mean climate state, does the climate variability have some impact on land-atmosphere coupling? A theoretical study found that the strength of the external forcing can affect the coupling strength and the location of coupling hot spots (Wei et al. 2006). Even less is known about how the land-atmosphere coupling is related to the different timescales of climate variability. The intraseasonal variability of precipitation has a strong influence on the soil moisture variability (Wei et al. 2008), but little has been done on the connection between this variability and land-atmosphere coupling.

In this paper, we reviewed our recent work on the impact of atmospheric variability on soil moisture-precipitation coupling, mainly from Wei et al. (2010b) and Wei and Dirmeyer (2010). The paper first presents our results of GLACE-type experiments with two different Atmospheric General Circulation Models (AGCMs) coupled to three different land surface schemes (LSSs). The large-scale connections between precipitation predictability, land-atmosphere coupling strength, and climate variability are examined, and the roles of different model components and different action processes in land-atmosphere coupling are investigated. Based on the analyses, the model estimated land-atmosphere coupling strength can be calibrated to account for errors in the simulation of precipitation variability, a quantity that is observable in the large scale and found to be closely related to the coupling strength.

2. Models and experiments

The two AGCMs are a recent version of the Center for Ocean-Land-Atmosphere Studies (COLA) AGCM (Misra et al., 2007; Kinter et al., 1997) and a recent operational version of the National Center for Environmental Prediction (NCEP) Global Forecast System (GFS) model. The COLA AGCM is configured with 28 vertical sigma levels, while GFS is configured with 64 vertical sigma levels. They both have a spectral triangular truncation of 62 waves (T62) in the horizontal resolution (approximately 1.9° grid). The three LSSs are: the latest version of the COLA simplified Simple Biosphere model (SSiB) (based on Xue et al., 1991; Dirmeyer and Zeng 1999), the version 3.5 of the Community Land Model (CLM3.5) (Oleson et al., 2004, 2008), and a recent version of the Noah land model (Ek et al., 2003). Wei et al. (2010a) gave a brief introduction of the recent changes of these LSSs. There are many specific differences among these LSSs in the parameterization of particular processes. In addition, the three LSSs have different numbers of soil layers and soil depths, and each uses its own soil and vegetation data sets.

Two experiments are preformed in this study:

1. GLACE-type experiments are performed with each of the six different model configurations. Detailed descriptions of the experiments and the indexes are in the Appendix. The ensemble W is the same as the standard GLACE experiment, while in ensemble S the soil moisture in all the soil layers is replaced, instead of only the subsurface soil moisture, in order to make the results from different LSSs comparable (see Appendix).
2. As the two AGCMs have different precipitation variabilities (shown below), which may lead to different soil moisture variabilities, the purpose of experiment (2) is to investigate the respective impacts of atmospheric variability and soil moisture variability on land-atmosphere coupling. Modified GLACE-type experiments are performed with COLA-SSiB and GFS-SSiB. The difference from experiment (1) is that, in the S runs, all members of the COLA-SSiB ensemble reads the same soil moisture from one W run of GFS-SSiB, while all members of the GFS-SSiB ensemble reads the same soil moisture from one W run of COLA-SSiB. The W ensembles are the same as in experiment (1). Although both from SSiB, the soil moisture climatologies of the two model configurations will be somewhat different, but this effect should be small compared to that of the dramatically different variabilities driven by precipitation.

3. Results from GLACE-type experiments

Fig. 1 shows the Ω_p values of total precipitation for ensembles W (16-member control experiment for June-July-August (JJA)) and S (soil wetness specified in all ensemble members from an arbitrarily chosen member of W) and their difference $\Omega_p(S)-\Omega_p(W)$ (see the appendix for complete definitions). The three indexes are generally higher when the LSSs are coupled to the COLA AGCM than to GFS, indicating that the difference in AGCM is the main reason for these differences. The impact of different LSSs, which can be seen from the varying spatial distributions of the indexes when coupled to the same AGCM, is secondary. Ω_p shows largely similar patterns for all the six model configurations, with the largest values in the tropical rain belt where the SST forcing has the strongest influence (Shukla 1998). The patterns of $\Omega_p(W)$ and $\Omega_p(S)$ are very similar, with large differences ($\Omega_p(S)-\Omega_p(W)$) mainly over the regions with common high values. This indicates that the land-atmosphere coupling strength may be strongly influenced by the external forcing. By "external", we mean the forcing is from outside of the land-atmosphere system, such as that from SST. The patterns of $\Omega_p(S)-\Omega_p(W)$ for different model configurations have much lower similarity than those of Ω_p (spatial correlations are 0-0.29 for $\Omega_p(S)-\Omega_p(W)$ and 0.43-0.71 for Ω_p). For both AGCMs, coupling to SSiB produces the strongest land-atmosphere coupling strength globally, while coupling to Noah produces the weakest. The differences seen should be mainly from the land models' different connections between soil moisture and surface fluxes, because they are coupled to the same AGCM.

4. A decomposed view of land-atmosphere coupling strength

As discussed above, the slowly varying boundary forcing may play an important role in the similarity of the precipitation time series in different ensemble members (magnitude of Ω_p). It is very likely that the "fingerprints" of these slow forcings also exist in the precipitation time series. An effective way to examine this is to decompose the time series by frequency bands. After ignoring the first 8 days of integration of each JJA to avoid possible problems associated with the initial shock to the model atmosphere, as in calculating Ω_p, there remain 84 days left for analysis. We performed a discrete Fourier transform (discussed in detail in Ruane and Roads (2007)) to decompose the daily time series into three frequency bands: fast synoptic (2-6 days), slow synoptic (6-20 days), and intraseasonal (20-84 days). The choice of these frequency bands is arbitrary; other comparable choices give similar results. Note that the time series may contain a portion of the seasonal cycle, but due to the length of the time series we refer the 20-84 days variation as intraseasonal.

Fig. 2 shows the variance percentages of precipitation in these three bands for model simulations and the observationally based Global Precipitation Climatology Project One-Degree Daily (GPCP-1DD) datasets (at 1°×1° resolution, from 1997-2009) (http://precip.gsfc.nasa.gov/gpcp_daily_comb.html; Huffman et al. 2001). For a specific AGCM, the three model configurations are very consistent in their variance distributions. However, compared to the GPCP-1DD data, all the model simulations underestimate the high-frequency (fast synoptic) variance and overestimate the low-frequency (intraseasonal) variance, especially over tropics and subtropics. Multi-year simulations of these models,

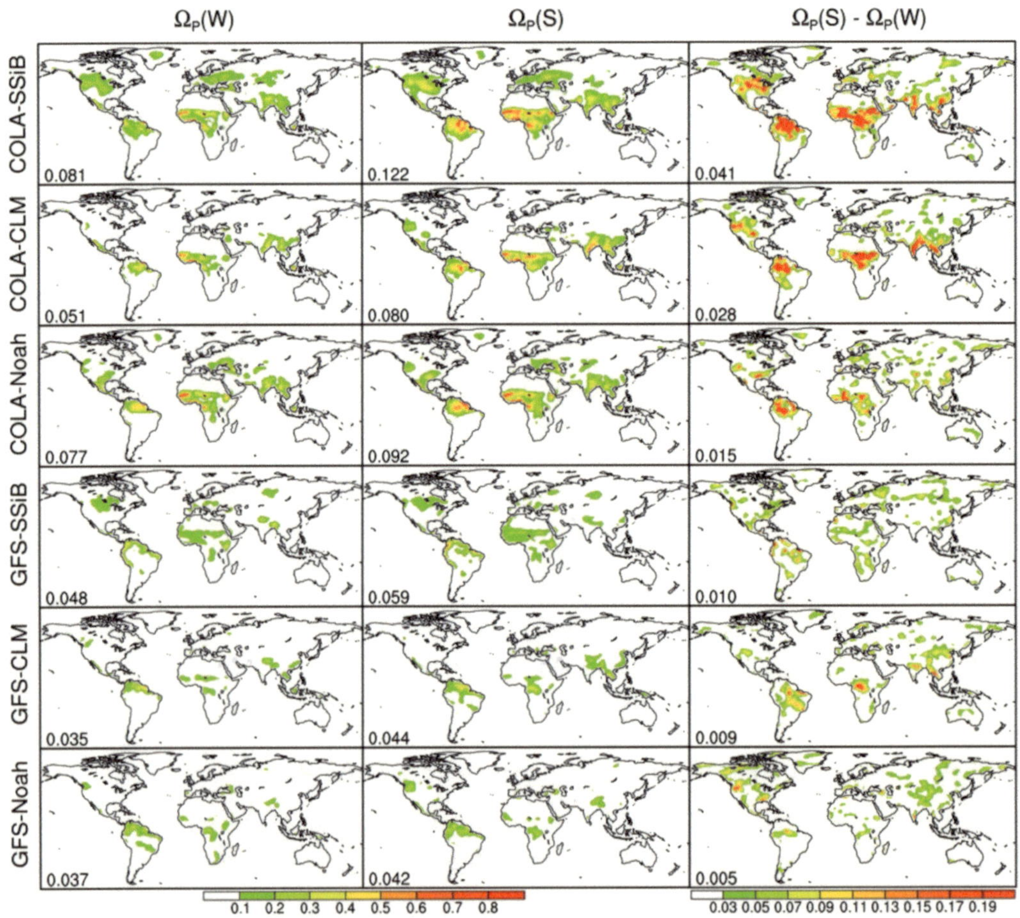

Fig. 1. The GLACE parameter Ω for precipitation from ensembles (left column) W and (middle column) S, and (right column) their difference. The six rows are for six different model configurations. The global mean (land only) value of each panel is shown at the left corner.

have similar variance percentage distributions as these GLACE-type simulations (not shown).

For theoretical white noise, the variance at each frequency is the same, so the variance percentages are determined by the widths of the frequency bands. Therefore, the variance percentages of the above three bands (from fast to slow) for white noise are: 69%, 21%, and 10%, respectively. Overall, both the model results and observations follow a red spectrum, with variance percentages less than white noise values at high frequencies and greater than white noise values at low frequencies.

In Fig. 2, the spatial correlations between $\Omega_p(W)$ and the percentage of intraseasonal variance (IV) are high (right column), but the correlations of $\Omega_p(W)$ with the other two frequency bands are negative (left two columns). Ensemble S (not shown) shows similar results as ensemble W. This demonstrates that regions with a larger percentage of IV tend to have a higher value of Ω_p, no matter whether soil moisture is interactive (W) or not (S).

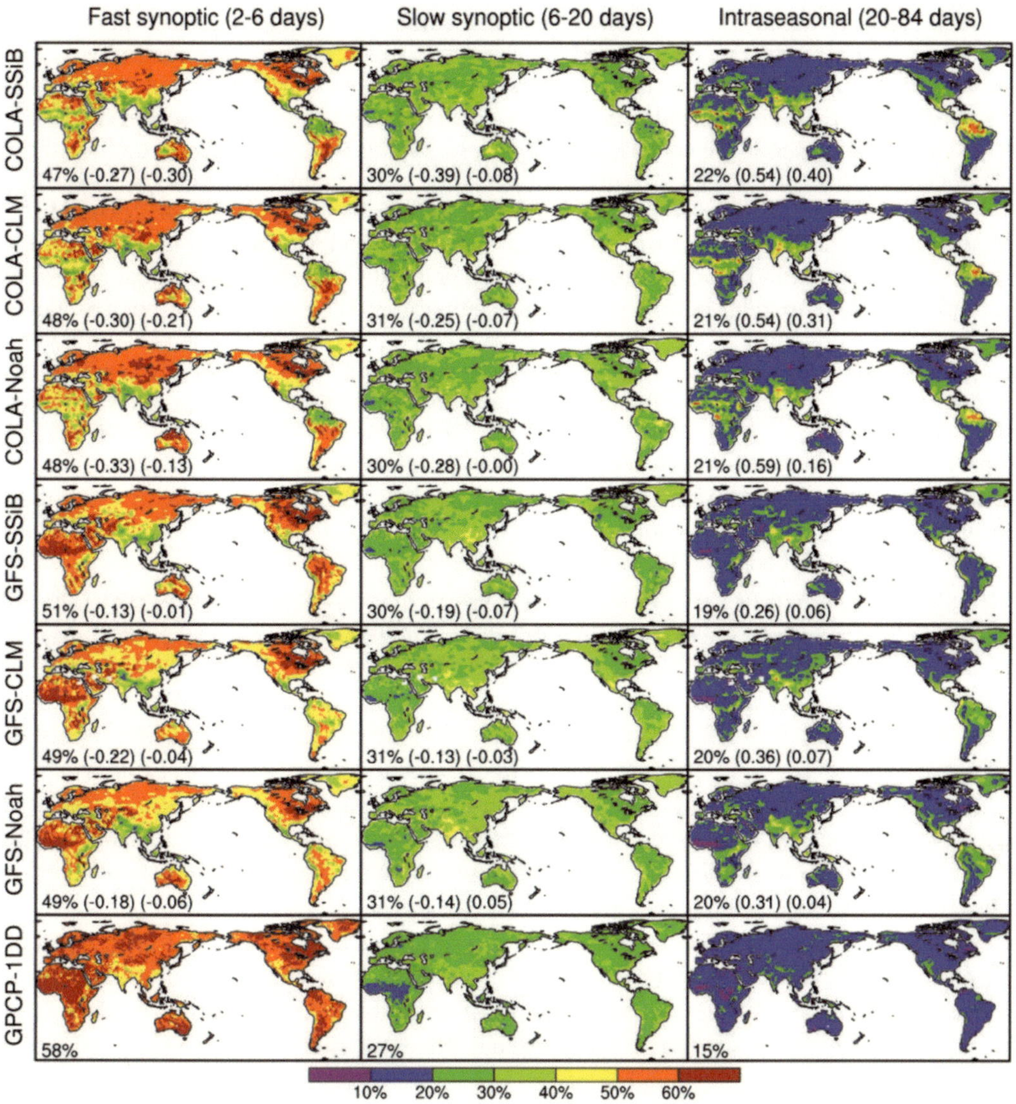

Fig. 2. The average variance percentages of JJA daily precipitation time series in three frequency bands: fast synoptic (2-6 days; left column), slow synoptic (6-20 days; middle column), and intraseasonal (20-84 days; right column). The top six rows are from six different model configurations (all from ensemble W; ensemble S has similar results), and the bottom row is from the observationally based dataset of GPCP-1DD. The value (or three values) at the left corner of each panel is the global mean percentage, (the spatial correlations of the variance percentage with $\Omega_p(W)$, and with $\Omega_p(S) - \Omega_p(W)$). The GPCP-1DD datasets are shown at 1°×1° grid; interpolating them to model grid does not affect the results.

This is not unexpected because, as we discussed above, most of the precipitation predictability (or Ω_p) is from the slowly varying boundary forcing. Regions with stronger boundary forcing may be constrained to show more low-frequency variation and the precipitation time series will be more similar in an ensemble (larger Ω_p). For ensemble S, the prescribed soil moisture is also one of the slow boundary forcings. However, compared to the ensemble without the constraint of this slow forcing (W), ensemble S does not show significant change in the global pattern of variance distribution (ensemble S does show overall less low-frequency variance and more high-frequency variance than ensemble W because of the lack of soil moisture interaction (Delworth and Manabe 1989)). These results show that different land models or land states do not matter much for the global pattern of precipitation variance distribution, which may be determined by other factors such as global climate (SST, radiation etc.) and the convection scheme. Ruane and Roads (2008) obtained similar results from a global assimilation system. They found that two different land models did not produce a noticeable difference in variance distribution of precipitation, but two different convection schemes can have significantly different effect. Wilcox and Donner (2007) also showed that the convection parameterization of a GCM can greatly impact the frequency distribution of rain rate, and their model with relaxed Arakawa-Schubert formulation of cumulus convection (also used in COLA AGCM) exhibit a strong bias toward excessive light rain events and too few heavy rain events.

The above shows that neither the land model nor soil moisture has a great impact on the global pattern of precipitation variability and predictability. However, their impact may be strong at regional scales. The difference $\Omega_p(S)-\Omega_p(W)$ shows the impact of soil moisture. It tries to remove the effects of the same strong external forcing on both S and W and highlight the role of soil moisture, although we understand that the effects of those forcing cannot be completely removed in a nonlinear system (more discussion on this aspect follows). The spatial correlations between percentage of IV and $\Omega_p(S)-\Omega_p(W)$ are also shown in Fig. 2 (as the last number in right column). They are generally weaker than the correlation with Ω_p, indicating that something other than the low-frequency external forcing is playing an important role in $\Omega_p(S)-\Omega_p(W)$. This should be the impact of soil moisture.

Wei et al. (2010b) demonstrates that the pattern of Ω_p and $\Omega_p(S)-\Omega_p(W)$ in Fig. 1 can be reproduced by using only the time series of intraseasonal precipitation variation (higher frequencies filtered out), and the time series of the other two frequency bands result in very weak values. This is because the intraseasonal component of precipitation, mostly caused by the same low-frequency external forcing, has high consistency among the ensemble members, while the high frequency component of precipitation, mostly from chaotic atmospheric dynamics, is generally incoherent among the ensemble members. This result indicates the importance of IV in the estimation of land-atmosphere coupling.

The overestimation of low-frequency variance shown in Fig. 2 is also consistent with the overestimation of precipitation persistence shown in Fig. 3. The lag-2-pentad autocorrelation of precipitation (ACR) shown here is also an indicator of the percentage of low-frequency precipitation variance, and it has similar spatial distributions as the percentage of IV but is much easier to calculate. More importantly, its spatial distributions are more similar to that of $\Omega_p(W)$ than the percentage of IV, as can be seen from the much higher spatial correlations (Fig. 3). This is probably because the percentage of IV only considers the variation at a certain frequency band (20-84 days here) but ACR considers the general persistence and is not restricted by certain frequencies. It can also be seen in Fig. 3 that the precipitation variability of GFS is overall closer to that of GPCP data (Xie et al. 2003) than the COLA AGCM, which may affect the accuracy of the simulated land-atmosphere coupling. This will be discussed next.

This model bias has also been shown in some other studies and by comparing with other observational datasets. Although the observational datasets have uncertainties and errors, Sun et al. (2006) found that no matter what observational dataset is used, this model bias is relatively large compared to the uncertainties among observations. This bias of the models may be related to a well-known problem in AGCM parameterizations: premature triggering of convection so that precipitation falls too frequently but too light in intensity (Trenberth et al. 2003; Sun et al. 2006; Ruane and Roads 2007).

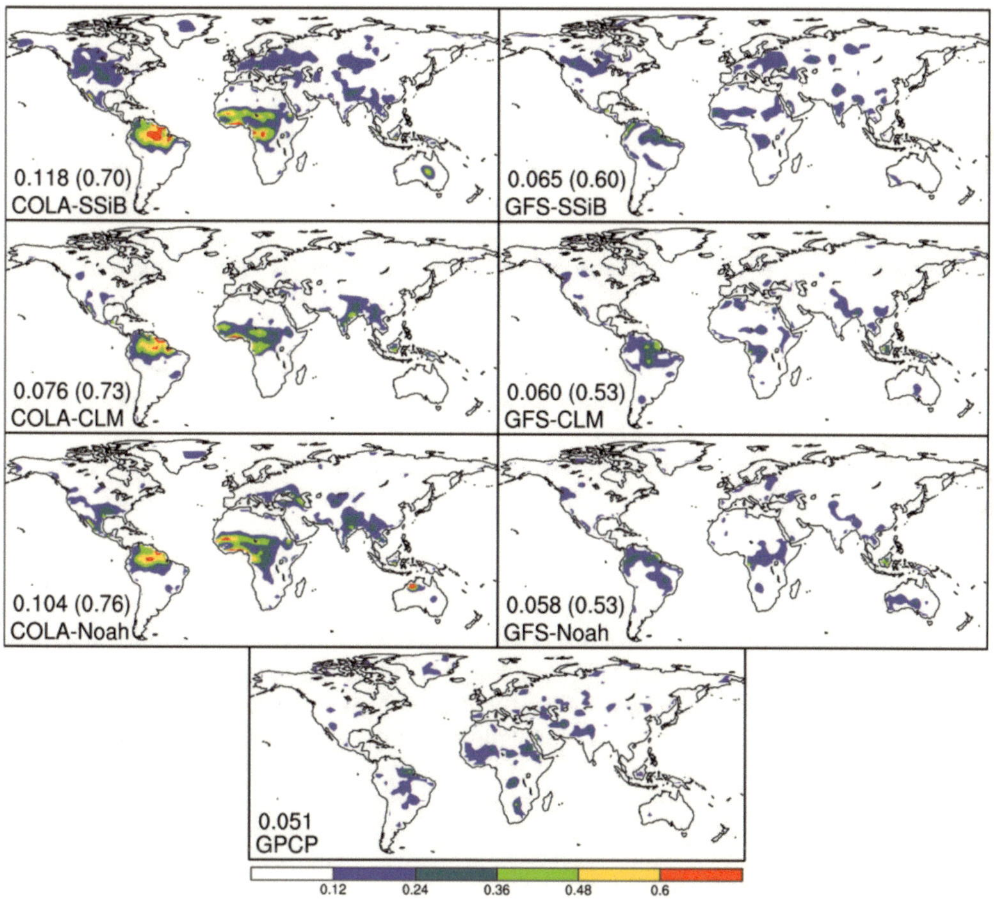

Fig. 3. The JJA lag-2-pentad autocorrelation (ACR) of pentad precipitation time series for (top six) different model configurations and (bottom) GPCP data. The model data are from 16 ensembles of W (sample size 16x16=256) , while the GPCP data are from16 years (1987-2002) to match the sample size of the models. Values larger than 0.12 are over 95% confidence level. Seasonal cycles are not removed in this calculation; removing them can lead to results with similar patterns but smaller amplitude.

5. Respective role of land and atmosphere in soil moisture-precipitation coupling

The above has shown that the low-frequency precipitation variability has an indirect but important connection to the computed land-atmosphere coupling. More low-frequency precipitation variability in the model can lead to higher precipitation predictability (Ω_p) and stronger land-atmosphere coupling ($\Omega_p(S)-\Omega_p(W)$). Therefore, there are three different processes involved: SM→ET, ET→P, and precipitation variability. What is the relationship among them? GUO06 separated SM→ET and ET→P based on a post hoc analysis, but they did not explicitly separate the role of soil moisture and atmosphere because ET is strongly affected by precipitation and radiation; the variability of ET is an approximation of low-frequency atmospheric variability. Thus SM→ET inevitably includes some information from atmosphere, including precipitation variability. The multi-model coupling approach provides a unique tool to estimate the respective impacts of the AGCMs and LSSs on the coupling, and only by this approach can the role of atmosphere and land be completely separated.

Although the above experiment shows the dominant role of the AGCMs in land-atmosphere coupling, it is still uncertain what are the roles of land and atmosphere in the coupling because the characteristics of the AGCMs may also affect land and its response. How important is the land response compared to the characteristics of the atmosphere (including atmospheric variability, sensitivity of precipitation to ET, etc.)? As the precipitation has more persistence in the COLA AGCM than in GFS, this attribute of precipitation variability is stored in the soil moisture, with more sustained soil states when the LSSs are coupled to the COLA AGCM than coupled to GFS (not shown). In order to investigate the impact of soil moisture variability on the coupling, in experiment (2) we exchange the prescribed soil moistures for COLA-SSiB and GFS-SSiB in ensemble S. This forces the models to see different soil moisture variabilities from their original S ensembles, and there is no change to the W ensembles. The resulting impacts on precipitation predictability (or coupling strengths) are shown in Fig. 4 (denoted $\Omega_P(S')-\Omega_P(W)$). It can be seen that, compared to the original coupling strength in Fig. 1, the modified coupling strength are overall weaker for COLA-SSiB and stronger for GFS-SSiB, but COLA-SSiB still has much stronger coupling strength than GFS-SSiB. This indicates that the impact of soil moisture variability may have some impact on the land-atmosphere coupling, but the characteristics of the atmosphere appear to be more important, at least for the case here. Note that the above action processes may be model dependent and vary spatially, but it is important to know that the atmospheric variability may also impact the coupling strength indirectly through land. Therefore, the precipitation variability impacts soil moisture-precipitation coupling both directly in the atmosphere and indirect through land (Fig. 5). More low-frequency variability of soil moisture usually means more sustained dry and wet periods and stronger low-frequency evaporation variation, which can lead to a more robust precipitation response (higher predictability and coupling strength). The direct impact of precipitation variability on soil moisture-precipitation coupling has been discussed in section 4 and more discussion follows.

GUO06 calculated SM→ET as $(\Omega_E(S)-\Omega_E(W))\cdot\sigma_E(W)$, where Ω_E is defined as in (A1) but for ET, and $\sigma_E(W)$ is the standard deviation of the 6-day average ET for the W runs. This definition considers two factors: a robust ET response to soil moisture ($\Omega_E(S)-\Omega_E(W)$) and

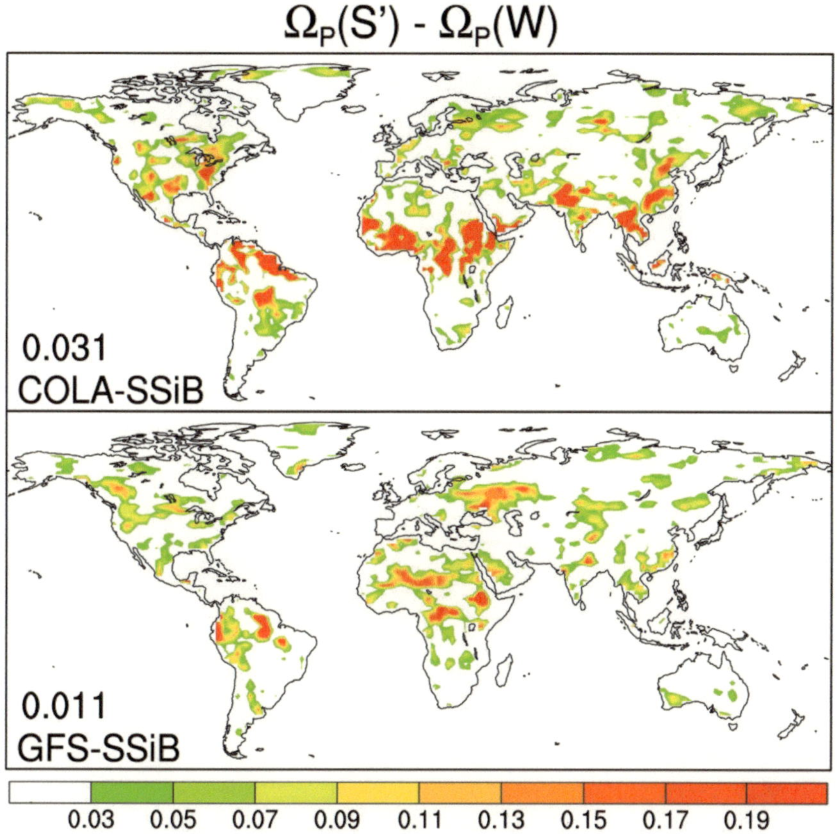

Fig. 4. Same as the right column of Fig. 1, but (top) the S runs of COLA-SSiB read soil moisture from a W run of GFS-SSiB, and (bottom) the S runs of GFS-SSiB read soil moisture from a W run of COLA-SSiB (from Wei and Dirmeyer 2010).

a high variability of ET ($\sigma_E(W)$). For soil moisture to have a strong impact on ET, both of them need to be sufficiently high. ET→P is simply calculated by GUO06 as the ratio of $\Omega_P(S)-\Omega_P(W)$ to SM→ET. (GUO06 introduced one more method for calculating ET→P, which produces similar results.) As mentioned, this diagnostic of SM→ET should be affected by the variability of precipitation and radiation. The experiment (2) above also demonstrates this indirectly. To verify this in another way, we calculate the inter-model correlation between ACR and SM→ET across the 12 GLACE models (Koster et al., 2006) (a correlation with a sample size of 12). We show results from GLACE models because we do not want our results to be limited to the models we use. It can be seen in Fig. 6a that there is substantial positive correlations between ACR and SM→ET over the globe, supporting our conjecture on the relationship between precipitation variability and SM→ET. The correlations between ACR and ET→P and between SM→ET and ET→P are both very low (Fig. 6b, 6c), suggesting that they are largely independent.

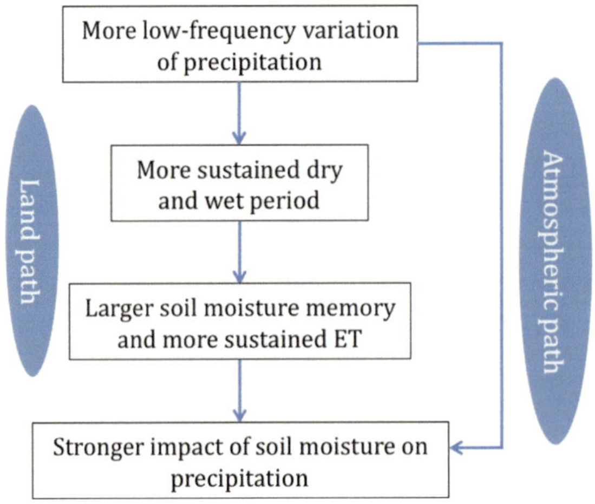

Fig. 5. Schematic of the impact of precipitation variability on soil moisture-precipitation coupling.

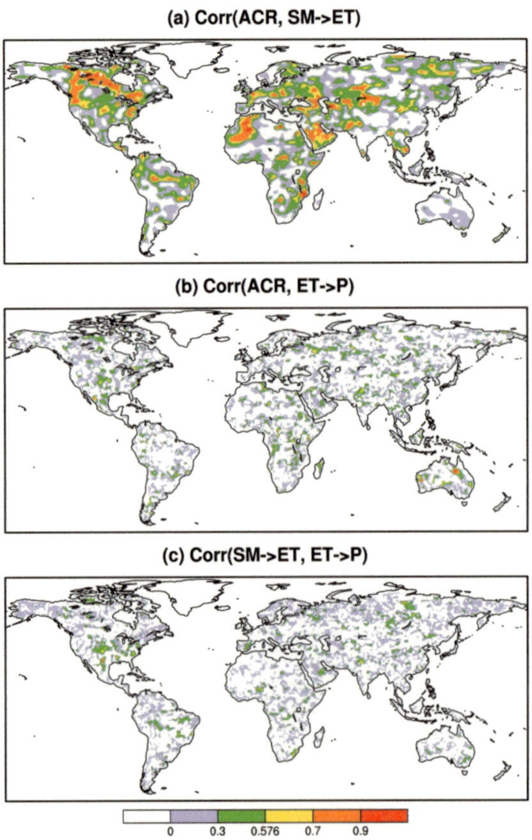

Fig. 6. The correlations between (a) ACR and SM→ET, (b) ACR and ET→P, and (c) SM→ET and ET→P across the 12 models participating in GLACE. The values over 0.576 are significant at the 95% level (from Wei and Dirmeyer 2010).

6. Conceptual relationships

The above analysis shows that the spatial distribution of both $\Omega_P(W)$ and $\Omega_P(S)$ are largely consistent with that of the low-frequency variability of the atmosphere, which may come from the slow external forcing or internal atmospheric dynamics. We denote it as F. F can be measured by the percentage of IV or ACR, and we have shown above that ACR is a better metric. The conceptual relationship between Ω_p , F, and the impact of soil moisture α is given as

$$\Omega_p = F \cdot (\alpha_0 + \alpha) \tag{1}$$

where α_0 is a constant, and α_0 >> α over most regions. Thus, the spatial variation of Ω_p is largely determined by F, which is consistent with the analysis above. F is similar for both ensemble W and ensemble S, so the coupling strength

$$\Omega_p(S) - \Omega_p(W) = F \cdot (\alpha(S) - \alpha(W)), \tag{2}$$

where $\alpha(S) - \alpha(W)$ is the difference of α between the two ensembles and can be further expanded to SM→ET and ET→P. Therefore

$$\Omega_P(S) - \Omega_P(W) = F \cdot SM \rightarrow ET(F) \cdot ET \rightarrow P, \tag{3}$$

where SM→ET is a function of F and some other model parameterizations. All the three factors—F, SM→ET, and ET→P may impact the coupling strength. The impact of F is separated from that of SM→ET because the impact of the atmosphere can be independent of the land surface. This multiplicative form of the equation considers the nonlinear combination of the factors. When SST is prescribed, F is mainly a property of the AGCM, especially the convection scheme. SM→ET is affected by both the LSS and the AGCM, and ET→P is mainly determined by the AGCM, especially the convection and boundary layer parameterizations. This decomposition, although is still conceptual, integrates our current understanding on land-atmosphere coupling, and it makes diagnosing land-atmosphere coupling much easier.

GUO06 only partly considers the impact of F (through SM→ET) and attributes the rest of the coupling strength to ET→P. They found that, for the 12 GLACE models, SM→ET has stronger correlation with the coupling strength than ET→P, and concluded that SM→ET is the main cause of the differences in the coupling strength. However, we show that the differences in SM→ET can be partly attributed to the impact of atmospheric variability, so it is still hard to say whether the different AGCMs or the different LSSs is the main cause of the differences in coupling strength. In spite of that, for our six model configurations here, the multi-model coupling method has clearly shown that the difference between the AGCMs is the main reason. It remains possible that the differences among the three LSSs are unusually small or the differences between the two AGCMs are unusually large.

7. Calibration of the estimated GLACE land-atmosphere coupling strength

In order to examine whether our results on the overestimation of low-frequency variance and its relationship with Ω_p also apply to other models, we look at the GLACE dataset

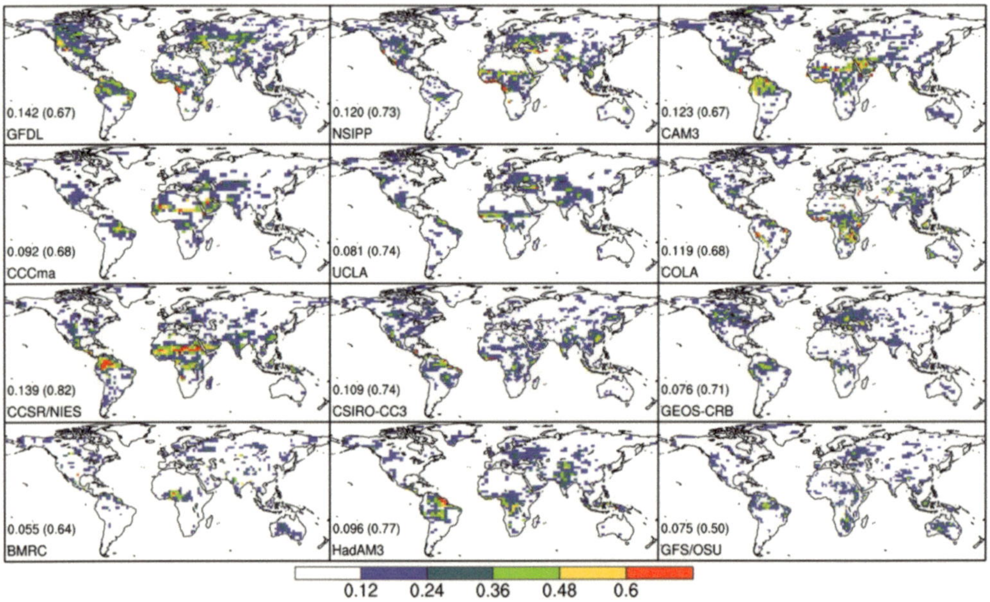

Fig. 7. Same as Fig. 3, but for the 12 models participating in GLACE (all from ensemble W; ensemble S has similar results). The first value at the left corner of each panel is the global mean (land only), and the second value (in the parentheses) is the spatial correlation of ACR with $\Omega_p(W)$.

(Koster et al. 2006). Fig. 7 shows the ACR for 12 models participating in the GLACE, and their respective spatial correlations with Ω_p. Similar to our model simulations above, ensemble S (not shown) shows very similar results to ensemble W. Also, all the models here have overestimated the mean ACR compared to the GPCP results, and their average is about double the ACR of GPCP (Fig. 8). The spatial correlations of ACR and Ω_p are always high; even the lowest value (0.5 from GFS/OSU) is well over the 99% confidence level (assume the grid points are independent). Therefore, the GLACE models and our models show similar relationships between Ω_p and ACR.

We have shown that the estimate of precipitation predictability caused by soil moisture ($\Omega_P(S)-\Omega_P(W)$) is closely related to the atmospheric low-frequency variability F but the models generally overestimate it. The influence of F on $\Omega_P(S)-\Omega_P(W)$ is obviously shown in equations (2) and (3). However, we cannot conclude that the land-atmosphere coupling strength estimated by GLACE is overestimated, because other important factors (SM→ET and ET→P) are still not observed at large scale. Nonetheless, we may assume that the other factors from the model ensemble are better than that of most individual models, and try to correct F to make $\Omega_P(S)-\Omega_P(W)$ possibly closer to reality. Roughly, we calibrate the average $\Omega_P(S)-\Omega_P(W)$ for the 12 models at each grid point (all interpolated to a common 2.5°×2.5° grid as GPCP data):

$$(\Omega_P(S)-\Omega_P(W))_{calibrated}=(\Omega_P(S)-\Omega_P(W))\frac{ACR(\text{obs})}{ACR(\text{models})}, \tag{4}$$

where ACR(obs)/ACR(models) is the ratio of ACR for GPCP data and the average ACR for the 12 models. This method tries to correct F by scaling it with a ratio of observed and modeled precipitation ACR. The GPCP-1DD data is used as ground truth observations. The ACR is calculated from ensemble W instead of ensemble S because it corresponds more closely to the real world (soil moisture is interactive). Note that the precipitation time series also contains a slice of the seasonal cycle during JJA, which is consistent with GLACE analysis. The predictability from seasonal variation is also important, because not every model can produce an accurate seasonal cycle. This calibration method considers model biases in both intraseasonal and seasonal variances. If removing the predictability from the seasonal variation, the coupling strength will be weaker but the patterns are similar (not shown).

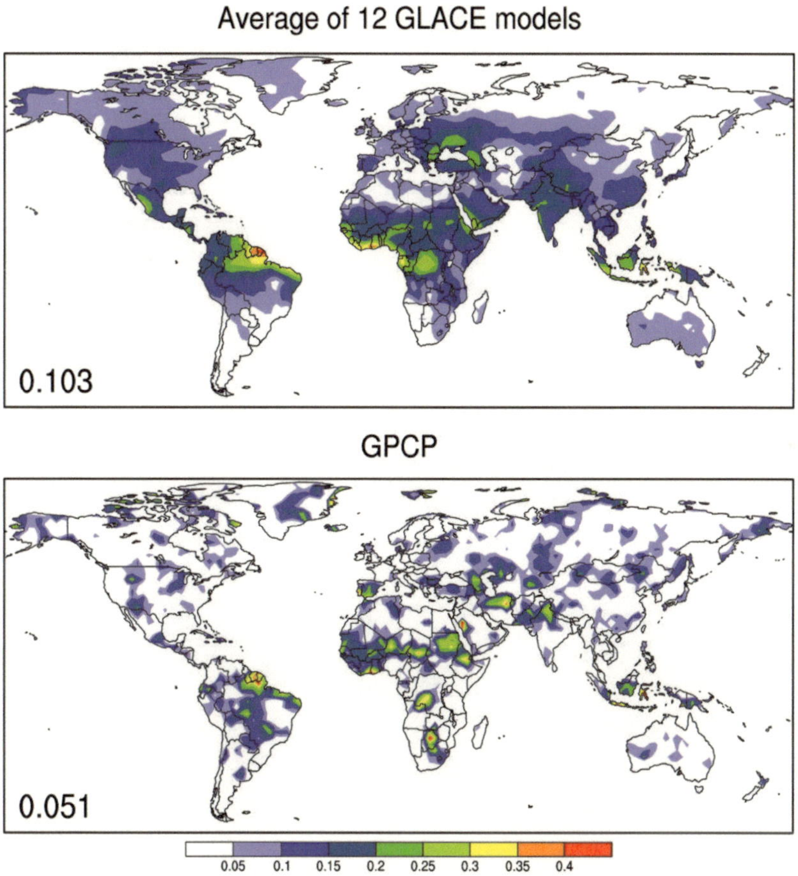

Fig. 8. Same as Fig. 3, but (top) for the average of 12 models participating in GLACE and (bottom) GPCP (same as in Fig. 3).

Although the spatial correlations between Ω_p and the ACR are very significant for all the models, the strong connection between Ω_p and F described in equation (1) may not happen over all the regions. The noise may damp the connection in some cases. We then calculated the correlation between Ω_p and ACR across the 12 GLACE models (a correlation with a sample size of 12; Fig. 9). It indicates a general connection between Ω_p and ACR for all the 12 models at each grid point. The regions with strong positive correlation are where Ω_p and ACR have a strong connection for almost all the models. Over other regions where the correlation is positive but not so strong, their relationship may not be so consistent for the models but larger F can generally lead to larger Ω_p. We can see in Fig. 9 that over 95% of the land areas show positive correlations and more than half land areas show significant correlations (at 95% level), which supports our assumptions on their relationship.

Based on the above analysis, we perform a calibration over the regions where the correlation in Fig. 9 is over 95% confidence level (0.576), using equation (4). The results are shown in Fig. 10. It can be seen that the coupling strength reduces by about 20% after calibration, but global pattern is similar to the original one. The coupling strength is significantly weakened over US Great Plains, Mexico, and Nigeria. The pattern over India and Pakistan changes a little. In Wei et al. (2010b), we have preformed a similar calibration using the percentage of IV instead of ACR. The results are largely similar, but the coupling strength over US Great Plains weakened less.

Due to the changing and heterogeneous nature of the relationship between Ω_p and F, our calibration method is not flawless. However, the results illustrate how the amplitude and distribution of coupling strength may change after some rectification of the model bias. The unique design of GLACE makes its results difficult to evaluate by directly comparing them with observational variables, even if these large-scale observations exist, because the GLACE metric is based on ensemble statistics, and observations present us with only one "ensemble member". Some recent studies using observational based data have cast doubt on the strong coupling strength in the Great Plains (Ruiz-Barradas and Nigam 2005, 2006; Zhang et al. 2008), but whether their results are comparable to the GLACE result need further study. On the other hand, Wang et al. (2007) have shown that less restrictive metrics than measuring ensemble coherence, such as the change in overall precipitation variance between S and W cases, reveals even more areas of apparently strong coupling strength.

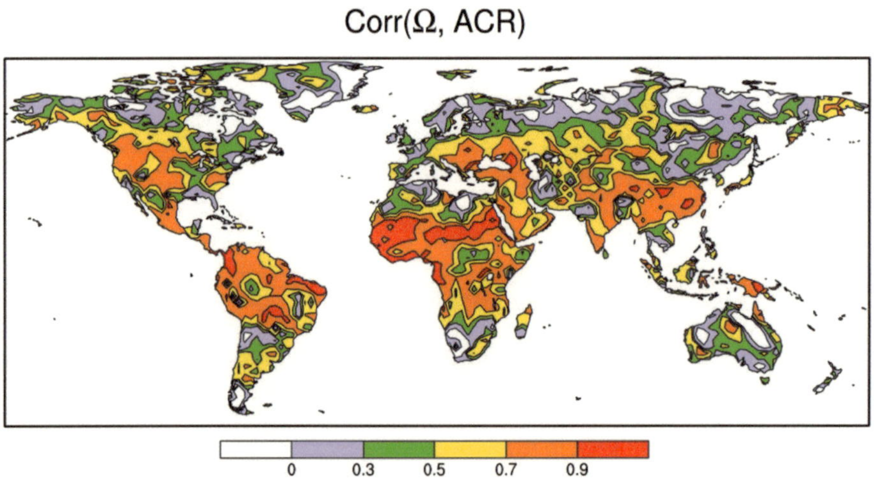

Fig. 9. The correlation between $\Omega_P(W)$ and ACR across the 12 models participating in GLACE.

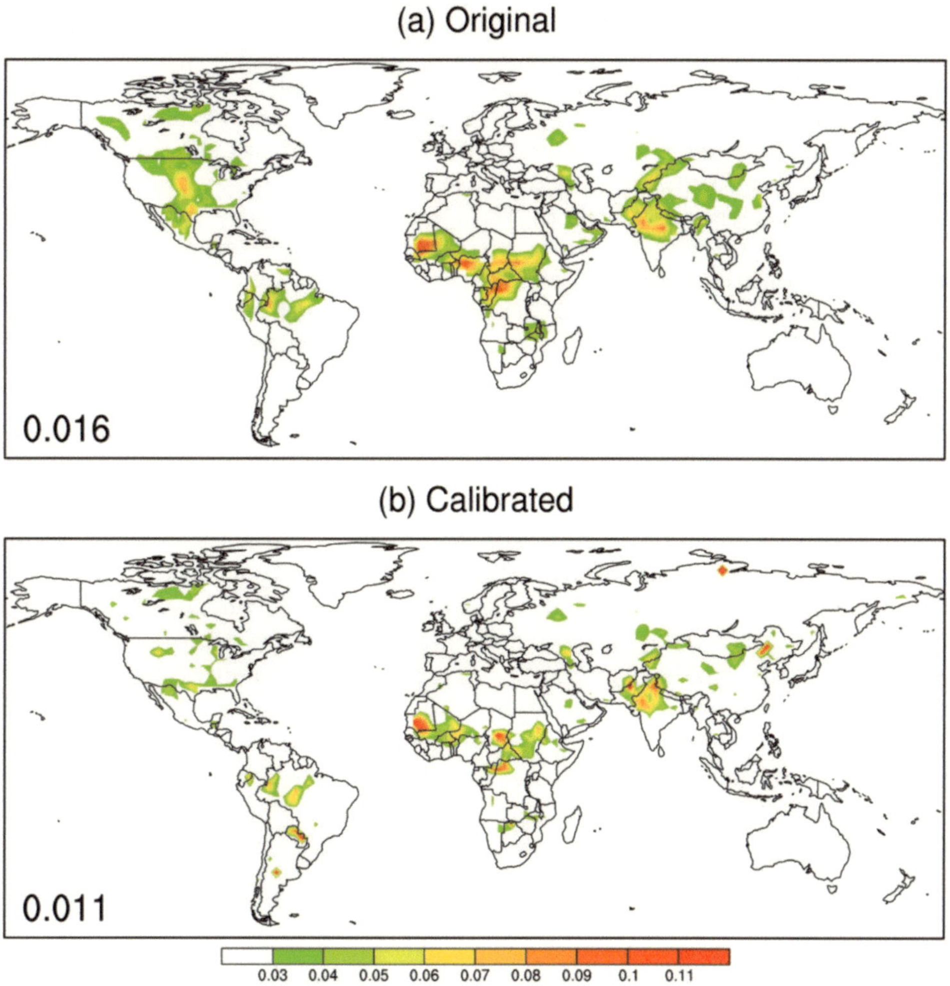

Fig. 10. The estimated land-atmosphere coupling strength ($\Omega_P(S)-\Omega_P(W)$). (a) From the original GLACE dataset. (b) Calibrated result. Only the regions where correlation in Fig. 8 exceeds 0.576 (95% confidence level) are calibrated. The global average value is shown at the left corner of each panel.

8. Conclusions and discussion

Coupling one AGCM to three different land models gives us an unprecedented opportunity to study the role of different components in land-atmosphere interaction. The behavior of the coupled models and their land-atmosphere interaction are investigated by a set of GLACE-type experiments. It is found that the two AGCMs determine the overall spatial distribution and amplitude of precipitation variability, predictability, and land-atmosphere coupling for the six model configurations. The impact of different LSSs is mostly regional. Different LSSs or soil moisture have little influence on the global pattern of precipitation predictability and variance distribution because of the stronger control of other factors. The estimated precipitation predictability and land-atmosphere coupling strength is closely related to the low-frequency variability of atmosphere, which can impact land-atmosphere coupling both directly in the atmosphere and indirectly through soil moisture response to precipitation. Based on these findings, the land-atmosphere coupling strength is conceptually decomposed into the impact of low-frequency precipitation variability, the impact of soil moisture on evapotranspiration (ET), and the impact of ET on precipitation. As most models participating in GLACE have overestimated the low-frequency component of precipitation, a rough calibration to the GLACE-estimated land-atmosphere coupling strength is performed. The calibrated coupling strength shows a similar global pattern, but is significantly weaker over some regions, like US Great Plains and Mexico.

We discussed the land-atmosphere coupling strength based on the index defined in GLACE, which emphasizes the temporal coherency of the precipitation time series among the ensemble members. Analysis with another index defined by Wang et al. (2007), which emphasizes the relative divergence of mean precipitation, does not show significant overall differences in the coupling strength between GFS and COLA AGCM configurations. This suggests the important role of atmospheric variability in determining the different GLACE coupling strengths of GFS and COLA AGCM configurations.

The integration period of these GLACE-type experiments is JJA, so the longest timescale in this study is intraseasonal. The effect of land may be different for longer timescales. Our judgment on the connections between external forcing, low-frequency precipitation variability, and Ω_p and $\Omega_P(S)-\Omega_P(W)$ is based on model results so may not be absolutely true in reality. Some intraseasonal variations of precipitation may not come from external forcing but from the flow instabilities (especially in midlatitudes; e.g., Charney and Shukla 1981) that are not fully understood nor properly simulated. The observational datasets used in this study are all based on satellite observations, and they may have consistent biases. In spite of these limitations, this study qualitatively separates the role of external forcing and local soil moisture on precipitation variability and predictability, and increases our understanding of land-atmosphere interaction.

9. Appendix

9.1 Global Land-Atmosphere Coupling Experiment (GLACE)

GLACE (Koster et al. 2004, 2006) is a model inter-comparison study focusing on evaluation of the role of land state in numerical weather and climate predictions. It consists of sets of 16-member ensembles of AGCM experiment (we only discuss two sets here). Ensemble W as a set of free runs with different initial land and atmosphere conditions but forced by the same SST from 1994, and ensemble S is the same as ensemble W except that, at each time step, the soil moisture in all the soil layers is replaced by that from one member chosen from ensemble W (all members of S have the same soil moisture). This is a little different from that of the standard GLACE experiments, where only subsurface soil moisture was replaced in the S ensemble. We design the experiments in this way to make the results from different LSSs more comparable; it has been shown that the upper layer of Noah model is responsible for an unusually large part of evapotranspiration (Zhang et al. 2010). All runs cover the period of 1 June-31 August, 1994. A diagnostic variable Ω was defined in GLACE:

$$\Omega = \frac{16\sigma^2_{<X>} - \sigma^2_X}{15\sigma^2_X}, \qquad \text{(A1)}$$

where $\sigma^2{}_X$ is the intraensemble variance of variable X, and $\sigma^2{}_{<X>}$ is the corresponding variance of ensemble mean time series. In calculating the variance, the first 8 days of data of each run is discarded to avoid model initial shock, and the remaining 84 days are aggregated into 14 six-day totals. Therefore, $\sigma^2{}_X$ is a variance across 224 (16×14) six-day totals from all the ensemble members, and $\sigma^2{}_{<X>}$ is a variance across 14 six-day totals from the ensemble mean time series.

Theoretically, if the 16 members of an ensemble have identical time series of X, $\sigma^2{}_{<X>}$ will be equal to $\sigma^2{}_X$ and Ω will be 1; if the X time series of the 16 members are completely independent, $\sigma^2{}_{<X>}$ will be equal to $\sigma^2{}_X / 16$ and Ω will be 0. Without sampling error, Ω will range between 0 and 1. Ω measures the similarity (or predictability) of the time series in 16 ensemble members. Analyses show that Ω emphasizes the temporal coherency, or the phase relationship, more than the mean and temporal variance of the time series (Wang et al. 2007; Yamada et al. 2007). Mathematically, Ω is equivalent to the percentage of variance caused by the slowly varying oceanic, radiative, and land surface processes (Koster et al. 2006; Yamada et al. 2007). The difference of Ω from the two ensembles, $\Omega(S)-\Omega(W)$, is then equivalent to the percentage of variance caused by the prescribed soil moisture, and is a measure of land-atmosphere coupling strength in GLACE.

10. Acknowledgement

This research was supported by National Oceanic and Atmospheric Administration award NA06OAR4310067. The computing was completed on NCAR supercomputers. We thank all the model groups participating in GLACE for providing their experimental results.

11. References

Charney, J. G. and J. Shukla, 1981: Predictability of monsoons. Monsoon Dynamics, Editors: Sir James Lighthill and R. P. Pearce, Cambridge University Press, pp. 99-109.

Delworth, Thomas, Syukuro Manabe, 1989: The Influence of Soil Wetness on Near-Surface Atmospheric Variability. J. Climate, 2, 1447–1462.

Dirmeyer, P. A., 2006: The hydrologic feedback pathway for land-climate coupling. J. Hydrometeor., 7, 857-867.

Dirmeyer, P. A., and F. J. Zeng, 1999: An update to the distribution and treatment of vegetation and soil properties in SSiB. COLA Technical Report 78, Center for Ocean-Land-Atmosphere Studies, Calverton, MD, 27 pp.

Dirmeyer, P. A., R. D. Koster, and Z. Guo, 2006: Do global models properly represent the feedback between land and atmosphere? J. Hydrometeor., 7:1177–1198.

Dirmeyer, P. A., Z. Guo, and J. Wei, 2010: Building the case for (or against) land-driven climate predictability. iLEAPS Newsletter. No. 9, 14-17.

Dirmeyer, P.A., M.J. Fennessy, and L. Marx, 2003: Low Skill in Dynamical Prediction of Boreal Summer Climate: Grounds for Looking beyond Sea Surface Temperature. J. Climate, 16, 995–1002.

Ek, M. B., and Coauthors, 2003: Implementation of Noah land surface model advances in the National Centers for Environmental Prediction operational mesoscale Eta model, J. Geophys. Res., 108(D22), 8851, doi:10.1029/2002JD003296.

Gao, X. and P. A. Dirmeyer, 2006: A multimodel analysis, validation, and transferability study of global soil wetness products. J. Hydrometeor., 7:1218–1236.

Graham, N.E., T.P. Barnett, R. Wilde, M. Ponater, and S. Schubert, 1994: On the Roles of Tropical and Midlatitude SSTs in Forcing Interannual to Interdecadal Variability in the Winter Northern Hemisphere Circulation. J. Climate, 7, 1416–1441.

Guo, Z., and Coauthors, 2006: GLACE: The Global Land–Atmosphere Coupling Experiment. Part II: Analysis. J. Hydrometeor., 7, 611–625.

Guo, Z., P. A. Dirmeyer, X. Gao, and M. Zhao, 2007: Improving the quality of simulated soil moisture with a multi-model ensemble approach. Quart J.Roy. Meteor. Soc., 133, 731-747.

Huffman, G.J., R.F. Adler, M. Morrissey, D.T. Bolvin, S. Curtis, R. Joyce, B McGavock, J. Susskind, 2001: Global Precipitation at One-Degree Daily Resolution from Multi-Satellite Observations. J. Hydrometeorol., 2, 36-50.

Joyce, R. J., J. E. Janowiak, P. A. Arkin, and P. Xie, 2004: CMORPH: A method that produces global precipitation estimates from passive microwave and infrared data at high spatial and temporal resolution. J. Hydrometeorol., 5, 487-503.

Kinter J. L., and Coauthors, 1997: The COLA atmosphere-biosphere general circulation model. Vol. 1: Formulation. COLA Tech. Rep. 51, 46 pp. Center for Ocean–Land–Atmosphere Studies, Calverton, MD.

Koster R. D. and Suarez M. J., 1995: Relative contributions of land and ocean processes to precipitation variability. Journal of Geophysical Research 100, 13775–13790.

Koster, R. D., and Coauthors 2004: Regions of strong coupling between soil moisture and precipitation, Science, 305, 1138-1140.

Koster, R. D., and Coauthors, 2006: GLACE: The Global Land-Atmosphere Coupling Experiment. Part I: Overview, J. Hydrometeorol., 7, 590–610.

Koster, R.D., M.J. Suarez, and M. Heiser, 2000: Variance and Predictability of Precipitation at Seasonal-to-Interannual Timescales. J. Hydrometeor., 1, 26–46.

McPhee, J., and S.A. Margulis, 2005: Validation and Error Characterization of the GPCP-1DD Precipitation Product over the Contiguous United States. J. Hydrometeor., 6, 441–459.

Misra, V., and Coauthors, 2007: Validating and understanding ENSO simulation in two coupled climate models, Tellus, Ser. A, 59, 292–308.

Pitman, A. J. 2003: The evolution of, and revolution in, land surface schemes designed for climate models, Int. J. Climatol., 23, 479-510.

Quan, X.W., P.J. Webster, A.M. Moore, and H.R. Chang, 2004: Seasonality in SST-Forced Atmospheric Short-Term Climate Predictability. J. Climate, 17, 3090–3108.

Ruane, A. C., and J. O. Roads, 2008: Diurnal to Annual Precipitation Sensitivity to Convective and Land Surface Schemes. Earth Interactions, 12, 1–13.

Ruane, A.C., and J.O. Roads, 2007: 6-Hour to 1-Year Variance of Five Global Precipitation Sets. Earth Interactions, 11, 1–29.

Ruiz-Barradas, A., and S. Nigam, 2005: Warm Season Rainfall Variability over the U.S. Great Plains in Observations, NCEP and ERA-40 Reanalyses, and NCAR and NASA Atmospheric Model Simulations. J. Climate, 18, 1808–1830.

Ruiz-Barradas, A., and S. Nigam, 2006: Great Plains Hydroclimate Variability: The View from North American Regional Reanalysis. J. Climate, 19, 3004–3010.

Shukla J., and Coauthors 2000: Dynamical seasonal prediction. Bull. Amer. Meteor. Soc., 81, 2593–2606.

Shukla, J. 1993: Predictability of short-term climate variations. Prediction of Interannual Climate Variations. NATO ASI Series I: Global Environmental Change, Vol. 6, Editor: J. Shukla, 217-232.

Shukla, J. 1998: Predictability in the midst of Chaos: A scientific basis for climate forecasting. Science, 282, 728-731.

Su, F., Y. Hong, and D.P. Lettenmaier, 2008: Evaluation of TRMM Multisatellite Precipitation Analysis (TMPA) and Its Utility in Hydrologic Prediction in the La Plata Basin. J. Hydrometeor., 9, 622–640.

Sun, Y., S. Solomon, A. Dai, and R.W. Portmann, 2006: How Often Does It Rain? J. Climate, 19, 916–934.

Trenberth, K. E., A. Dai, R. M. Rasmussen, and D. B. Parsons, 2003: The Changing Character of Precipitation. *Bull. Amer. Meteor. Soc.*, 84, 1205–1217.

Wang, G., Y. Kim and D. Wang, 2007: Quantifying the strength of soil moisture-precipitation coupling and its sensitivity to changes in surface water budget. J. Hydrometeor., 8, 551-570.

Wei, J. and P. A. Dirmeyer, 2010: Toward understanding the large-scale land-atmosphere coupling in the models: Roles of different processes, Geophys. Res. Lett., 37, L19707, doi:10.1029/2010GL044769.

Wei, J., P. A. Dirmeyer, and Z. Guo, 2010b: How much do different land models matter for climate simulation? Part II: A decomposed view of land-atmosphere coupling strength. J. Climate. 23, 3135-3145.

Wei, J., P. A. Dirmeyer, Z. Guo, L. Zhang, and V. Misra, 2010a: How much do different land models matter for climate simulation? Part I: Climatology and variability. J. Climate. 23, 3120-3134.

Wei, J., R. E. Dickinson, and N. Zeng, 2006: Climate variability in a simple model of warm climate land-atmosphere interaction, J. Geophys. Res., 111, G03009, doi:10.1029/2005JG000096.

Wei, J., R.E. Dickinson, and H. Chen, 2008: A Negative Soil Moisture-Precipitation Relationship and Its Causes. J. Hydrometeor., 9, 1364–1376.

Wilcox, E.M., and L.J. Donner, 2007: The Frequency of Extreme Rain Events in Satellite Rain-Rate Estimates and an Atmospheric General Circulation Model. J. Climate, 20, 53–69.

Xie, P., J.E. Janowiak, P.A. Arkin, R. Adler, A. Gruber, R. Ferraro, G.J. Huffman, and S. Curtis, 2003: GPCP Pentad Precipitation Analyses: An Experimental Dataset Based on Gauge Observations and Satellite Estimates. J. Climate, 16, 2197–2214.

Xue, Y., P. J. Sellers, J. L. Kinter, J. Shukla, 1991: A simplified biosphere model for global climate studies, J. Climate, 4, 345-364.

Yamada, T.J., R.D. Koster, S. Kanae, and T. Oki, 2007: Estimation of Predictability with a Newly Derived Index to Quantify Similarity among Ensemble Members. Mon. Wea. Rev., 135, 2674–2687.

Zhang, J., W.-C. Wang, and J. Wei, 2008: Assessing land-atmosphere coupling using soil moisture from the Global Land Data Assimilation System and observational precipitation, J. Geophys. Res., 113, D17119, doi:10.1029/2008JD009807.

Zhang, L., P. A. Dirmeyer, J. Wei, Z. Guo and C.-H. Lu, 2011: Land-atmosphere Coupling Strength in the Global Forecast System. J. Hydrometeor., 12, 147–156.

Part 2

Climate and Solar Activity

3

Solar Activity, Space Weather and the Earth's Climate

Maxim Ogurtsov[1], Markus Lindholm[2] and Risto Jalkanen[2]
[1]Ioffe Physico-Technical Institute
[2]Finnish Forest Research Institute
[1]Russia
[2]Finland

1. Introduction

The Sun is the ultimate source of energy for the Earth. Energy radiating from our closest star provides the natural power that fuels most of the physical and biological processes important to life. The radius of the Sun is 6.96×10^{10} cm and angular diameter ≅1919″ (arc seconds). Average distance between the Sun and the Earth is 1.496×10^{13} cm (1 astronomical unit). The mass of the Sun is 1.99×10^{33} g, and luminosity - 3.84×10^{26} W. Solar atmosphere consist of the photosphere and the chromosphere. The **photosphere** is the zone from which the sunlight we see is emitted. It is few hundreds of kilometers thick and has a temperature of 5500–6000° C. The **chromosphere** is an irregular stratum above the photosphere where the temperature rises from about 6000° C to about 20000° C. The chromosphere is 2000–3000 km thick and the colorful image of this layer can be observed during solar eclipses.
Solar activity is a set of non-stationary processes and phenomena in the Sun's atmosphere associated with the changes in solar magnetic fields. Manifestations of solar activity include the emergence and further time evolution of sunspots, faculae, flocculae, protuberances and coronal loops, solar flares and fluxes of solar wind and electromagnetic radiations. *Sunspots* are the most prominent observable signs of solar activity. They are temporary phenomena on the photosphere that appear visibly as dark spots contrasting with surrounding area. They are caused by rather intense magnetic fields, which inhibit convection, forming areas of reduced surface temperature. The diameter of a typical sunspot is 22000–29000 km (30–40″). The corresponding area is 120–160 millionths of the visible hemisphere or Micro Solar Hemisphere (MSH). Very large sunspots can reach diameters of 60000 km or more while the smallest sunspots are roughly 3500 km in diameter (Solanki, 2003). Each sunspot is characterized by a dark core, the umbra, and a less dark halo, the penumbra. Umbra to photosphere brightness is about 0.24, and penumbra to photosphere brightness is about 0.77. The magnetic field strength in the photosphere is approximately 1000–1500 G averaged over a sunspot. The effective temperature of a sunspot is 4200 K (de Jager, 2005). Mean lifetime of a sunspot is of an order of 10 days while some of them can exist up to one half a year. The complex of several sunspots forms a *sunspot group*. The readily observable rise and fall of sunspot number over an approximate 11 year period between minima is the most prominent feature of sunspot activity - the well-known cycle of Schwabe. Modern heliophysics deals with both statistical (synthetic) and physical indices of solar activity.

Statistical indices are determined from data of instrumental observations of the Sun using special mathematical algorithms. They have no direct physical meaning. Physical indices reflect the actual physical solar phenomena and quantify directly measurable manifestations of solar activity (such as radio flux and UV flux). Sunspot number is the most widespread statistical solar index. An astronomer from the Zürich observatory, Rudolf Wolf introduced the relative sunspot number in the middle of the 19th century. The sunspot series initiated by him is called the Zürich or Wolf sunspot number R_Z and it is still widely used as a statistical measure of solar activity. This series starts in AD 1700 (Fig. 1A). Recently Hoyt and Schatten (1998) performed a thorough and widespread archive search and substantially increased the amount of original information. They introduced a new index of solar activity called the group sunspot numbers R_G. This data set spans the time interval of AD 1610–1995.
Sunspot area measurements are available from the Greenwich Observatory for the period 1874–1976. The Greenwich series is based on daily photographic images of the Sun. Recently, this series was substantially expanded by Nagovitsyn et al. (2004) who used additional information obtained by Schwabe, Spörer and de la Rue before 1874 as well as Soviet-Russian astronomers after year 1976. The sunspot area can be considered a physical index, since it is directly linked to the solar magnetic flux emerging at sunspots (Nagovitsyn, 2005).
Faculae are long-lived (typical lifetime is 3 times longer than that of sunspots) bright areas, usually situated near sunspots. They occur on the photosphere, but sometimes can extend upwards into the chromosphere. The temperature of faculae is about 300^0 C higher than that of background photosphere, and magnetic field is close to 400 G (Obridko, 2008). Their brightness is usually 1.10–1.45 of the photosphere brightness.
Luminosity or the *total solar irradiance* (TSI - often referred to as the solar constant) is another important index of solar activity. TSI is the wavelength-integrated intensity of solar electromagnetic radiation. Solar constant has been measured since 1978 by satellite radiometers. Two composite TSI records - ACRIM () Fig. 1E) and PMOD series (Fröhlich and Lean, 1998)- were calculated from the original satellite data.
A *solar flare* is a large explosion in the Sun's chromosphere. The energy released during a flare is typically on the order of 10^{27} erg s^{-1}. Solar flares can emit up to 10^{32} ergs of energy in various forms including accelerated particles and radiation emitted in spectral ranges from radio waves to X-rays and gamma rays.
The Earth has a mean radius of 6.37×10^8 cm and the area of the Earth's surface is 5.1×10^{18} cm^2. Oceans and seas cover 71% of the Earth's surface, i.e. 3.62×10^{18} cm^2. Terrestrial atmosphere consists of the troposphere (up to 10–17 km), the stratosphere (10–17 to 50 km), the mesosphere (50 - 80 km) and the thermosphere (above 80 km).
Magnetosphere is a spatial region where the motion of charged particles is governed by the Earth's magnetic field. It has a complex structure, determined by interaction between the magnetic field of solar wind, the Earth's magnetic field and the upper atmosphere plasma. Its base lies at the altitude of few hundreds of kilometers while its tail has a diameter of about 40 Earth's radii.
Geomagnetic activity is the perturbation of a geomagnetic field caused by changes in magnetosphere-ionosphere current system and closely connected to variations of solar particle flux. Geomagnetic storms, substorms and auroral phenomena are the main manifestations of geomagnetic activity. A strong magnetic storm can change the global field by 5×10^{-7} T or more. Kp index of geomagnetic activity is shown in Fig. 1D.

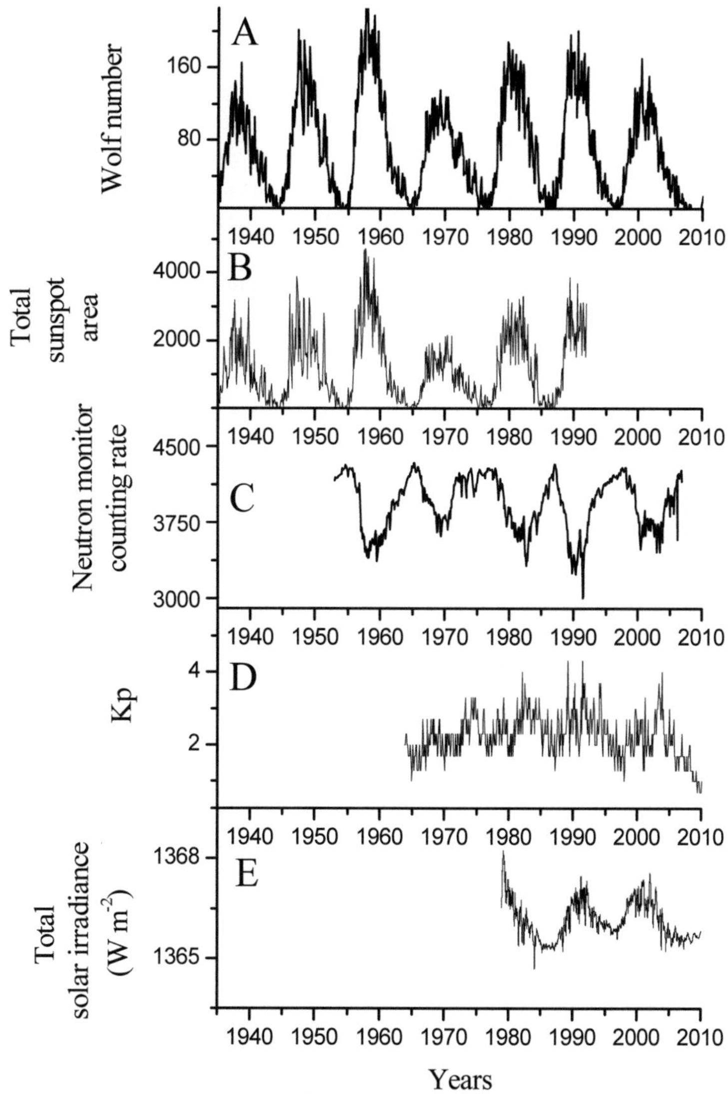

Fig. 1. Some important solar and geomagnetic indices. A – Wolf number, B – total sunspot area, C – counting rate of Climax neutron monitor (R_c=3 GV), D – geomagnetic K_p index, E – ACRIM record of the total solar irradiance. All plotted records are monthly mean data. They were taken from the sites: ftp://ftp.ngdc.noaa.gov/STP/SOLAR_DATA/; http://www.acrim.com/; http://www.gao.spb.ru/database/.

Solar-terrestrial physics or *heliogeophysics* comprises a number of scientific disciplines which study various phenomena and processes in the Sun and their manifestations in circumterraneous space and the Earth's atmosphere and magnetosphere. It is obviously a complex and extensive field, which includes solar physics, astrophysics, cosmic physics, geomagnetism, atmospheric physics, meteorology, climatology and, possibly, some other scientific spheres. Probable influence of solar activity on terrestrial climate is studied by *helioclimatology*. Climate is conveniently defined by means and terms of the weather.

Atmospheric weather is a physical state of the atmosphere in a certain place on the Earth at a given time or time interval, which is described by a set of meteorological parameters. Meteorological parameters are the features of the atmosphere's state including temperature, pressure, speed and direction of wind, quality and concentration of thermodynamically active admixtures (water droplets, vapor and aerosols). The changes of weather are the fluctuations of meteorological parameters connected with pure atmospheric processes.

Climate is a statistics of weather, which is characterized by meteorological parameters averaged over time intervals generally longer than 30 years.

Space weather is a physical state of the circumterraneous space (upper atmosphere, magnetosphere) at a given moment or time interval, which is described by a set of heliogeophysical parameters. The intensity of solar electromagnetic radiation and fluxes of solar cosmic ray (SCR), rate of ion generation, density and velocity of the solar wind particles, intensity of geomagnetic fluctuations are the main heliogeophysical parameters which form space weather. Space weather parameters averaged over long time intervals determine the *space climate*. Both space weather and space climate are closely connected with solar activity.

One of the enduring puzzles of climatology is how changes in the number of sunspots affect weather and climate on the Earth. A link between darkening of solar disk and rye prices was mentioned in fragments of Cato the Elder (234–149 BC) (Chizhevsky, 1973). B. Balliani was probably the first to assume a link between solar activity and terrestrial climate. In his letter to Galileo, Balliani noted that sunspots can be regarded as coolers of the Sun and, hence, of the Earth. Since this time search for possible solar-climate link has attracted the attention of many researchers. W. Hershel in 1801 revealed appreciable relationship between prices of some farm products, the harvest of which are weather dependent, and sunspot numbers. The work of Hershel was the starting point for a serious Sun-climate research. G. Wild (1882) likely was the first who noted effect of solar flares and geomagnetic disturbances on the surface temperature. In 1959 Ney suggested that intensity of GCR is an agent transferring solar influence on climate (Ney, 1959). Dickinson (1975) reported that fluxes of cosmic ray might influence cloudiness. Recent decades were a period of active investigation of a wide range of solar, solar-terrestrial and solar-heliosphere processes, both theoretical and experimental, performed using both groundbased and space-borne experiments. However, a lot of questions still remain and many problems have to be solved. Some of these questions are: What is the physical mechanism providing a link between solar activity, space weather and climate? Do variations in solar activity and GCR intensity contribute to global warming? What is the role of volcanic activity in long-term climatic change? Do change in geomagnetic dipole field affect terrestrial climate? The answer to these and many other questions require detailed knowledge about the history of climate, solar activity, space weather, volcanic activity and geomagnetic field over as long time scale as possible.

Unfortunately, the modern knowledge about the past of these phenomena is quite poor and has substantial gaps. The direct temperature records usually cover no more than last 100–150 years. The series of measurements of different parameters of solar activity also are short. The longest of them - group sunspot numbers - starts since AD 1610. Accurate observations of solar flares, research of GCR by means of neutron monitors and direct satellite measurements of different space weather parameters cover less than the last 55 years. Thus our current knowledge about many important heliogeophysical processes (the fluxes of solar and galactic cosmic rays, the solar wind velocity, the strength of interplanetary magnetic field, some geomagnetic indices) cover no more than 3–5 quasi eleven-year solar cycles. It is obvious that targeted observations of the Sun using various instruments cannot supply us this information. The necessary data can only be obtained by means of proxies provided by solar paleoastrophysics and paleoclimatology. Solar paleoastrophysics is the science which makes it possible to reconstruct different parameters of the Sun's activity in pre-instrumental era. Paleoclimatology is a scientific discipline focusing on climate's past. New achievements of paleoastrophysics and paleoclimatology obtained during the last 20–30 years allow us to investigate the evolution of solar activity and climate and examine their possible interrelations over the long time scales otherwise inaccessible to us.

2. Solar paleoastrophysics: advances and limitations

Paleoastrophysics is concerned with astrophysical phenomena whose signals reached the Earth before the time of instrumental astronomy. Solar paleoastrophysics uses both the data of historical chronicles (the catalogues of sunspot and aurorae naked eye observation) and indirect indicators of solar activity (the concentration of cosmogenic isotope and nitrate in natural archives).

2.1 Paleoastrophysics of cosmogenic isotope

The study of concentrations of cosmogeneous isotopes in natural archives is a one of the basic methods of solar paleoastrophysics. Cosmogenic radiocarbon ^{14}C and radioberyllium ^{10}Be originate in the Earth's stratosphere and troposphere due to the effect of energetic *galactic cosmic rays* (GCR). The GCR are charged particles with energies from about 1 MeV up to at least 10^{20} MeV. The source of GCR is outside the solar system but within the galaxy, most likely it is shock acceleration of super-nova remnants. GCR are observed with background neutron monitors (Fig. 1C), which are maximally sensitive to particles with energy of several GeV. The intensity of the GCR particles is given approximately by:

$$\frac{dN}{dE} \approx E^{-\gamma}\ , \mathrm{cm^{-2}\ s^{-1}\ GeV^{-1}\ sr^{-1}}, \tag{1}$$

where $\gamma \cong 2.6$ (E=10–10^6 GeV).

About 90% of the nuclei are hydrogen (protons), 10% helium (α-particles), and about 1% heavier elements (C, N, O). Electrons comprise about 1% of GCR (Bazilevskaya et al., 2008). In the inner heliosphere the low energy (<10^3 GeV) GCR are modulated by solar activity. Thus the intensity of the galactic cosmic radiation is highest during the solar cycle minimum and lowest during solar maximum. Present-day solar modulation of cosmic ray is produced by three major mechanisms (Jokipii, 1991):

a. The convection of the magnetic field and GCR outward caused by the solar wind flux.
b. The diffusion of GCR caused by their scattering by the irregularities in the interplanetary magnetic field (IMF). These irregularities are carried away from the Sun by the solar wind. The density of the heliospheric magnetic inhomogeneities depends on the Sun's activity level. When solar activity increases, the density of these heterogeneities rises. The solar wind velocity also shows some positive correlation with solar activity. Therefore the diffusion-convection effects cause the 11-year variation of

GCR intensity. The flux of cosmic ray particles with energy E=0.1–15 GeV during solar minimum is twice as large as during maximum phase (Stozhkov, 2003).

c. The drift modulation of GCR is caused mainly by changes of the polarity of the solar magnetic field. Positively charged cosmic rays preferentially enter the heliosphere from the direction of the solar poles during the time when the solar magnetic field in the northern hemisphere has negative polarity i. e. is directed outward (Jokipii, 1991). As a result, cosmic ray time dependence has a peaked form during solar cycles with negative polarity and it has a plateau form during cycles with positive polarity (see Fig. 1C). Since the Sun's magnetic field changes its polarity after every 11-year the cosmic ray intensity curve also appears to follow a 22-year cycle with alternate maxima being flat-topped and peaked. Theoretic calculations performed by Kocharov et al., (1995) has shown that during the global Maunder minimum the amplitude of drift variation can reach 15% for particles with E=0.5–50 GeV. Solar activity can also influence GCR intensity over a short time scale. Decrease of GCR intensity during several days after the large flares on the Sun and intensive solar coronal mass ejections is called a *Forbush decrease*. The amplitude of Forbush decrease (FD) typically is 2-5% for GCR with E=500 MeV. Powerful solar flares are another source of energetic particles in the Earth's vicinity. Energies of the *solar energetic particles* (SEP) can reach several tens of MeV and sometimes extend into the GeV range. The acceleration of particles by shock waves associated with coronal mass ejection can also play some role in SEP generation. More than 90% of solar energetic particles are protons. They can hit the Earth within 15 minutes to 2 hours after the solar flare. Energy spectrum of SEP can be described with formula:

$$\frac{dN(R)}{dR} \approx \exp(-R/R_0), \quad (2)$$

- where $N(R)$ is in cm^{-2} s^{-1} MV^{-1} sr^{-1}, $R=\frac{pc}{Q}$ is magnetic rigidity of particle in MV, p is the momentum, Q is the charge, and c is light velocity. R_0 is a characteristic rigidity, which is different for different flares. Typically R_0 lays in a range 50–100 MV, but for a very powerful flare of 23 February 1956 its value reached 325 MV. *Solar proton event* (SPE) is the enhancement of solar energetic particles in which proton flux with energy E_p>10 MeV is greater or equal to 10 proton flux units (pfu) or 10 particles cm^{-2} s^{-1} sr^{-1} (Kurt et al., 2004). SPE last from several hours to some days. Relativistic protons with energies up to 10-20 GeV are called (particularly in Russian scientific literature) *solar cosmic rays* (SCR). Fluxes of these particles can reach appreciable values and be observed with neutron monitors mounted on the ground. As a result *ground level enhancements* (GLE) occur.

Particles with energiy more than 1 GeV/nucleon are the main sources of radiocarbon generation. The major part of ^{14}C is produced by the secondary thermal neutrons in reaction $^{14}N(n,p)^{14}C$. The rate of atmospheric neutron production depends on the changes in cosmic ray flux. The mean rate of radiocarbon generation in the atmosphere is 2.2–2.5 atoms cm^{-2} s^{-1}or 6.5 kg of ^{14}C per year. The total mass of radiocarbon in the atmosphere is 45–75 tons. Calculations performed by Kocharov et al. (1990) showed that the contribution of SCR to the radiocarbon generation was only 10-15 % of the GCR contribution during 1956–1972. After origination ^{14}C is oxidized rapidly to ^{14}CO and then to $^{14}CO_2$, which, in turn, is homogenized in the atmospheric $^{12}CO_2$ pool and involved in a chain of geophysical and geochemical

processes forming the global carbon cycle, and is finally fixed by plants (e.g. tree rings). Radiocarbon decays with a half-life of 5730 years, which is enough in order to study processes which occurred up to 80 000 years ago.
Cosmogenic beryllium is generated in nuclear reactions $^{14}N(Ha,X)^{10}Be$, $^{16}O(Ha,X)^{10}Be$, where Ha are hadrons, X are the other reaction products. The reactions have a threshold character i.e. they take place only if the hadron energy exceeds 40–50 MeV. The mean rate of ^{10}Be generation in the atmosphere is (2.0–2.7) $\times10^{-2}$ at cm^{-2} s^{-1}. ^{10}Be oxidizes rapidly to ^{10}BeO, then it is captured by aerosols, washed out by precipitation, and preserved in polar ice and sea-bottom deposits. ^{10}Be decays with a half-life of 1.5□10^{6} years that is enough to investigate processes with time scales of few millions of years. Approximately two thirds of ^{10}Be and ^{14}C are produced in the stratosphere and the residual part is generated in the troposphere. The concentration of ^{14}C in tree rings is measured by proportional gas counters, liquid scintillation spectrometers and accelerator mass spectrometers. The measurements are expressed as $\Delta^{14}C$ which is the difference between the isotopically corrected activity of the sample and NBS standard ($[^{14}C]/[^{12}C]=1.176\times10^{-12}$). The dating of tree ring radiocarbon records is made by means of dendrochronology. Thus $\Delta^{14}C$ time series usually have 1 year time resolution. The concentration of ^{10}Be in ice is measured by accelerator mass spectrometry. The dating of beryllium series is made by recognizing layers of impurities attributed to known volcanic eruptions (volcanic markers) and by analysis of simultaneously measured records which have evident seasonal variations (H_2O_2, various ions).
Part of the incoming galactic radiation is deflected by the Earth's geomagnetic field. For this reason the production of cosmogenic isotopes is modulated by the changes in geomagnetic field. The concentration of radiocarbon in the atmosphere is modulated also by changes in global carbon cycle, particularly by variations in the rate of exchange between the atmosphere and mixed as well as deep layers of ocean. ^{10}Be abundance in polar ice depends on local meteorological conditions. Finally, the concentration of cosmogenic carbon and beryllium in natural archives is found to be dependent on the following factors: (a) GCR flux or spectral shape in the galactic vicinity of the solar system, (b) solar activity, (c) dipole geomagnetic field, (d) global and regional climate. It is important that cosmogenic nuclides ^{14}C and ^{10}Be provide a measure of GCR intensity, which is in turn effectively modulated by solar activity. Thus radiocarbon and beryllium records can be used for the reconstruction of solar activity variability over long time scales.
In the USSR, the radiocarbon studies of solar activity in the past began as early ago as 1965–1967 within the scope of the program Astrophysical Phenomena and Radiocarbon formulated by Konstantinov and Kocharov (1965). Eddy (1976) estimated how solar activity had varied during 5000 years using $\Delta^{14}C$ data set. He showed that the Sun has gone through both periods of very high solar activity (global maxima) and periods of very low solar activity (global minima). These works together with the efforts of J. Eddy, J. Schove, M. Stuiver, J. Beer and other investigators lay the foundations of solar paleoastrophysics. The longest radiocarbon series was measured in the framework of intercalibration program INTCAL98 (Stuiver et al., 1998) using rings of trees from Germany, Ireland and northwestern USA. The record covers the last 24 000 years and more than the last 10 000 years has decadal time resolution. This data set has been used for the longest sunspot number reconstructions (Fig. 2).

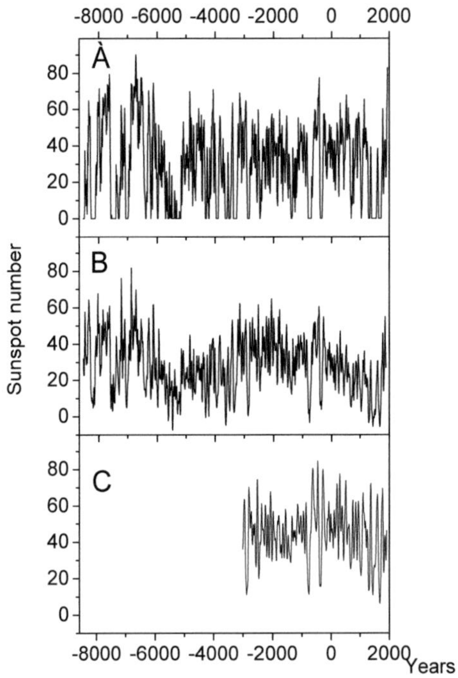

Fig. 2. Radiocarbon sunspot number reconstructions made by: A - Ogurtsov (2005); B - Solanki et al. (2004); C - Nagovitsyn et al. (2003). All the reconstructions were made using $\Delta^{14}C$ data of Stuiver et al. (1998).

The radiocarbon solar reconstructions show that the mean level value of sunspot number over the second part of 20th century (75-85 Wolf number units) was very high in the context of the last 10 000 years of sunspot history. Solanki et al. (2004) have concluded that the current episode of high solar activity since about the year 1940 is unique within the last 8000 years.

Annual solar reconstructions of different types over the last 1000 years has been produced (Fig. 3): reconstruction of Wolf number of Nagovitsyn (1997) who used the data of Schove (1983) based on historical accounts of aurorae; radioberyllium reconstructions of Usoskin et al. (2004) and Ogurtsov (2007) who used the long beryllium record from the South Pole (Bard et al., 2000); radiocarbon reconstructions of Nagovitsyn et al. (2003), Ogurtsov (2005), Solanki et al. (2004).

Fig. 3 shows that all the sunspot proxies agree in their major features. Coefficient of annual correlation between different reconstructions is 0.50–0.80. The correlation between reconstructed and instrumentally measured sunspot numbers reaches 0.70–0.80 over the decadal time scale. Thus the main global extremes of solar activity (Oort, Wolf, Spörer, and Maunder minimums and medieval and late medieval maximums) manifest themselves in all the sunspot paleoindicators. Establishing reliably the profound extremes of solar activity throughout the last millennium is an important achievement of solar paleoastrophysics. The analyses of the solar variability on the secular time scale is another field of paleoastrophysical research. The examination of long solar proxies made it possible to unequivocally prove the presence of the century-scale (55–135 yrs) variation of solar activity.

Recently it was shown that this periodicity - the cycle of Gleissberg - consists of two oscillation modes - 55–80 yrs variation and 90–135 yrs variation (Ogurtsov et al., 2002). Bi-centennial (170–260 yrs) solar variation - the cycle of Suess (de Vries) - was also discovered by means of analyzing the proxy data. Research on solar paleoindicators also gave serious evidence for the existence of 500-900-years and ca. 1500-years cycles of solar activity (Bond et al., 2001; Nagovitsyn, 1997; Ogurtsov 2010). Ca. 2300 year solar cycle - the cycle of Hallstatt - has been reported in a few works (see e.g. Vasiliev and Dergachev, 2002).

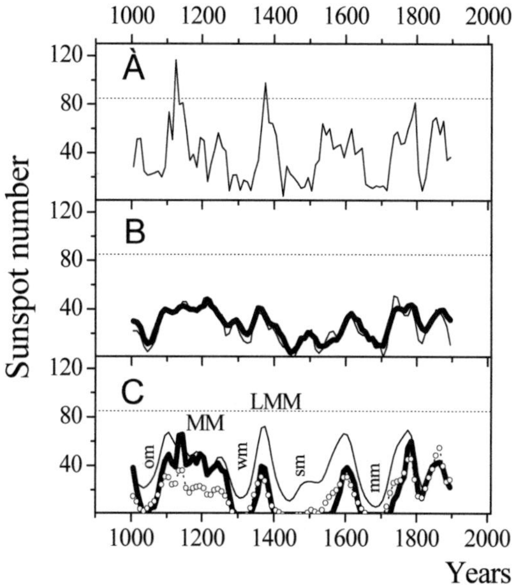

Fig. 3. Sunspot number: A - auroral reconstruction after Nagovitsyn (1997); B - beryllium reconstructions after Ogurtsov (2007, thin line) and Usoskin et al. (2004, thick line); C - radiocarbon reconstructions after Nagovitsyn et al. (2003, thin line), Ogurtsov (2005, thick line), Solankli et al. (2004, circles). MM - Medieval maximum, LMM - Late Medieval maximum, om - Oort minimum, wm - Wolf minimum, sm - Spörer minimum, mm - Maunder minimum.

In spite of many valuable results obtained by solar paleoastrophysics a number of problems still remain. Considerable uncertainty in our knowledge about: (a) the millennial-scale variability of geomagnetic dipole field and (b) past climatic changes as well as their influence on cosmogenic isotope abundance, currently challenges the precision of our paleoastrophysical understanding. For example, substantial difference between sunspot number reconstructions after Ogurtsov (2005) and Solanki et al. (2004) (see Fig. 1A,B) is caused mainly by different methods of long-term trend subtraction. Thus, Ogurtsov (2007) has concluded that in its present state sunspot paleoreconstructions most likely contain only qualitative information about the behavior of solar activity in the past but are not very suitable for extracting quantitative information. Nevertheless solar proxies let us elaborate some future scenarios of the evolution of solar activity. In the works of Ogurtsov (2005) and Solanki et al. (2004) it were shown that the average sunspot activity in the first part of 21 century most probably will be weaker than in the second part of 20th century.

2.2 Paleoastrophysics of nitrate

Nitrate ions (NO_3^-) concentration in polar ice of Antarctica and Greenland has been under investigation for many years (Zeller and Parker, 1981; Herron, 1982; Legrand et al., 1989; Mayewski et al., 1993; Dreschhoff and Zeller, 1994, 1998). The properties of nitrate record in ice are connected with its mechanism of generation. According to Logan (1983); Legrand and Kirschner (1990), Mayewski et al. (1990) nitrate "precursors" - the various NO_x (N, NO, NO_2) and NO_y (N, NO, NO_2, NO_3, HN_2O_5, N_2O_5, HO_2NO_2, $ClONO_2$, $BrONO_2$) molecules - are formed at different altitudes of the atmosphere:

a. In the troposphere (due to industrial activity, biomass burning, soil exhalation, lightning and the influence of galactic cosmic rays).
b. In the stratosphere and higher altitudes (due to biogenic N_2O oxidation, galactic cosmic rays, solar cosmic rays, solar UV radiation and relativistic electron precipitation).

Nitrate sources connected with cosmic radiation are the most important for paleoastrophysics. Energetic SCR and GCR particles as well as relativistic electrons precipitating from the radiation belts produces a lot of secondary free electrons with energies of hundreds of eV, which effectively interact with molecules of atmospheric gases forcing their ionization, dissociation and excitation:

$$O_2+e^-\rightarrow O(^3P)+O(^1D)+e^-,$$

$$N_2+e^-\rightarrow N^++N(^4S,^2D)+2e^-, \qquad (3)$$

$$N_2+e^-\rightarrow N(^4S)+N(^4S,^2D,^2P)+e^-.$$

The ions O_2^-, N^+, O^+, N_2^+, O_2^+, excited atoms of oxygen and nitrogen,□ are involved in a complex of photochemical reactions resulting in the generation of nitrogen oxide NO. It is presumed that each pair of ions creates 1.5 NO molecules. NO oxidizes to NO_2 within few tens of minutes while NO_2 lives 1-8 days and serves as a source for nitrate:

$$O_2^-+NO_2\rightarrow NO_2^-+O_2, \qquad (4)$$

$$NO_2^-+O_3\rightarrow NO_3^-+O_2. \qquad (5)$$

Reactions (4, 5) take place in the ozone layer in the stratosphere. Thus, a relationship between atmospheric ionization and production of nitrogen oxides appears. NO_3^- ions can generate clusters, particularly with water molecules:

$$NO_3^-HNO_3,\ NO_2^-(HNO_2)H_2O,\ NO_3^{\square}(H_2O)_n,\ n=2\text{–}5 \qquad (6).$$

These ion clusters have a long lifetime (up to 10^3–10^4 s) and thus can be fixed by aerosol particles. After fixation they precipitate on the Earth's surface both by gravitation sedimentation and downward air streams and become finally fixed in polar ice. Nitrate concentration in ice samples usually is measured by spectrophotometer.

Zeller and Parker (1981) as well as Dreschhoff and Zeller (1994, 1998) have reported the existence of an unequivocal link between SPE and short but prominent peaks in NO_3^- concentration both in Antarctica and Greenland. 11-year and 22-year cycles were also found by Zeller and Parker (1981), Dreschhoff et al. (1983) and Dreschhoff and Zeller (1998) in nitrate data. Significant but weak elevation in mean nitrate concentration after SPE was found in an Antarctic nitrate record by Palmer et al. (2001). Thus, analysis of nitrate records measured with ultra-high (few weeks to few months) time resolution can provide us with information about solar flare activity in the past. McCracken et al. (2001A) analyzed nitrate concentration in a 125.6 m long core drilled at Summit, Greenland (72^{0}N, 38^{0}W, altitude 3210 m) and two shorter cores drilled at Windless Bight (78^{0}S, 167^{0}E). The dating of these datasets was based on distinct seasonal variations in nitrate concentration and the data on known volcano eruptions. Because of the large input of sulfides into the atmosphere after each volcano eruption, the electrical conductivity in the ice layer of respective year increases sharply. The conductivity, measured along the ice core simultaneously with the nitrate concentration therefore provided scientists with a number of necessary time markers. McCracken et al. (2001A) identified 70 impulsive nitrate events between 1561 and 1950.

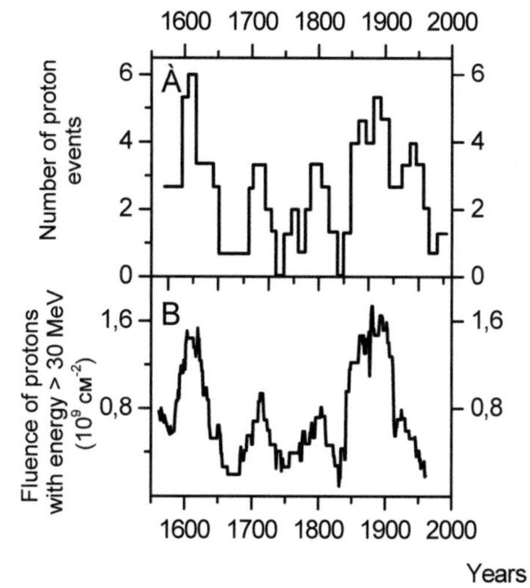

Fig. 4. A - the frequency of solar proton events averaged by a two solar cycle running means (scanned from McCracken et al. (2001B) and digitized); B - the fluence of powerful SPE reconstructed by McCracken et al. (2001A) averaged over 30 yrs.

The omnidirectional proton fluences (cm^{-2}) were also estimated. An analysis of this sequence performed by McCracken et al. (2001B) showed that solar proton events follow the century-scale (Gleissberg) periodicity during 1561-1950 (Fig.4). Interestingly the period of instrumental SCR measurement (after the mid-20th century) is a time of rather low frequency of SPE occurrence (see Fig. 4). It should be noted that some powerful SPE in the past were identified using other paleoindicators. Usoskin et al. (2006) discovered 10 SPE (1755, 1763, 1774, 1793, 1813, 1851, 1867, 1895, and 1927) by means of analysis of the data on concentration of ^{10}Be in Greenland ice. Kostantinov et al. (1992) performed joint analysis of the data on ^{14}C and ^{10}Be in natural archives. Substantial increases of SCR flux were established during 1750–1790, 1851–1853, 1868–1869, 1896. These results are in agreement with those of others (McCracken et al., 2001A) who identified SPE with strong (> 2×10^9 cm^{-2}) fluence of energetic (E_p>30 MeV) protons in 1755, 1763, 1774, 1793, 1813, 1851, 1866, 1868, 1895, 1896 and 1928.

Gleissberg periodicity was also revealed in the mean concentration of NO_3^- ions in Central Greenland ice (Kocharov et al., 2000). Mayewski et al. (1993) found a 112-year periodicity in their low time resolution nitrate series. Since fluence of solar particles with E>200-500 MeV is comparable with that of GCR particles, it is reasonable to assume the century-scale cyclicity in nitrate as the result of the corresponding variation in atmosphere ionization, which, in turn, is caused by the combined effect of the century-type variations in GCR and SCR fluxes. Quasi five-year variability is another important feature of nitrate concentration in Greenland ice (Kocharov et al., 1999). This quasi five-year variation is connected with a tendency of nitrate concentration to increase before and after sunspot maxima - at rise/decline phases of solar cycle (Kocharov et al., 1999; Ogurtsov et al., 2004). Such relationship may be a result of the corresponding tendency for strong SPE.

Despite the results which have confirmed that concentration of NO_3^- ions in polar ice actually is an indicator of stratospheric ionization and space weather, interpretations of the nitrate records still are controversial. Some researchers (Herron, 1982; Legrand and Kirchner, 1990) find no solar effects in nitrate data sets. Thus, these authors have rejected the possibility that nitrate concentration can reflect variations of solar activity and have proposed a weak contribution to the polar ice nitrate sequence from the middle atmosphere in comparison to the troposphere. An appreciable ambiguity still presently remains in our knowledge about transport of ions in atmosphere, deposital and post-deposital.
Since paleoastrophysical data have many uncertainties, the analyses of many proxies of different kinds are invaluable tools for the extraction of confident information.

3. Paleoclimatology: advances and limitations

Paleoclimatology reconstructs and studies climatic variations before the beginning of the instrumental measurements by means of different proxy indicators. Among the main data used by paleoclimatology are: tree rings (width and density), concentration of stable isotope (^{18}O, ^{13}C, D) in natural archives (ice, coral and tree rings) and historical documents.

3.1 Dendroclimatology

In 1892 P.N. Shvedov compared data on tree-ring growth with the data on precipitation of few weather stations from southeastern Russia. Shvedov (1892) concluded that droughts over the analyzed region have ca. 9 year cycle and emphasized the value of the tree-ring data for further climatic research (Dergachev, 1994). Later the relationship between radial tree growth and climatic factors was reported by Douglass (1914) as well as other scientists. A.E. Douglass is often considered as the father of dendroclimatology - the discipline, which estimates the past climate conditions from trees (dated tree rings mainly). The basic idea of dendroclimatology is that a discernible reaction occurs in growth increments of trees which grow under severe conditions due to variation of the limiting environmental factors. Trees living in extremely cold condition (northern tree line, upper tree line in mountains) reflect changes of summer temperature. Latewood density usually indicates temperature variation better than tree-ring width. Rings of trees growing in extremely dry conditions could serve as precipitation indicator. Tree rings are one of the most valuable climate proxies because they can be absolutely dated annually by means of dendrochronological cross-dating method. The works of Fritts (1976), Cook and Kairiukstis (1989) has helped dendroclimatology to become popular and over the last decades its methods have become major tools in reconstruction of past climates in many parts of the world (see e.g. Jones et al., 1998; Briffa, 2000; Briffa and Osborn, 2002; Mann and Hughes, 2002; Helama et al., 2005). Not only ring width responds to climate. Jalkanen and Tuovinen (2001) and Pensa et al. (2006) showed that in Lapland needle production of trees as well as annual tree height increment were strongly related to the mean air temperature of the previous summer. Thus tree-height and needle-trace chronologies provide a novel tool for reconstructing past summer temperatures. Tree-ring data usually are calibrated towards instrumental temperature. Although dendroreconstructions typically have their dating accurate to one year, their ability to encode long-term (centennial and slower) climate variability is often limited by the trend-removal technique applied to remove non-climatic variations in tree-ring time series. The longest temperature dendroreconstructions cover periods up to 7–8 millennia. Coefficient of correlation between instrumental temperature series and tree-ring reconstructions reaches 0.4–0.5 over annual time scale and more than 0.8 over decadal time scale. For latewood density reconstructions these values are 0.5–0.8 and over 0.9 respectively. Coefficient of annual correlation between tree-height increment proxy and measured temperature is 0.61.

3.2 Stable isotope climatology

In 1948 H. Urey calculated the temperature dependence of oxygen isotope fractionation between calcium carbonate and water and proposed that the isotopic composition of carbonates could be used as a paleothermometer (Urey, 1948). Later W. Dansgaard proposed the idea of using the isotopic composition of glacier ice as a climatic indicator (Dansgaard, 1954). Since isotopes with different masses have different rates of chemical reactions, natural geophysical and geochemical processes might cause isotope fractionation, which depends on temperature. As a consequence it is reasonable to expect the concentration of stable isotope ^{18}O, ^{13}C, D in natural archives - polar and mountain ice, bottom sediments, corals, tree ring cellulose - to hold information about past temperature. Isotope fractioning of the oxygen isotope has the following chain: ^{18}O is heavier than ^{16}O thus water vapor from the tropical ocean tends to have a slightly higher ratio of $^{16}O/^{18}O$ than that of the remaining ocean water. As water vapors traverses toward the poles, they lose the heavier, more easily condensed, ^{18}O water leading to lower and lower isotopic oxygen ratios. Consequently, the amount of ^{18}O relative to ^{16}O in the water vapor becomes less and less as it approaches the poles, preferentially losing ^{18}O water in the form of rain and snow. The ratio of oxygen isotope is determined by means of gas isotope-ratio mass spectroscopy. The measurements are expressed as $\delta^{18}O$ which is per mille (0.1 %) deviation from Vienna SMOW standard ($[^{18}O]/[^{16}O]=2005.2\times10^{-6}$). Stable isotope records have apparent seasonal cycle which can be used for annual dating accuracy. Deeper into the ice the annual layers become thinner and finally become indistinguishable. Therefore, the dating of the deeper layers depends considerably on the flow model. Discovery of Dansgaard-Oeschger events - rapid climate fluctuations from warm conditions (interstadials) to cold conditions (stadials) - is one of the most important successes of the paleoisotope thermometry. These events that occurred 25 times during the last glacial period (roughly 2-3 ka periodicity) were revealed by means of analysis of $\delta^{18}O$ concentration in the ice core retrieved from Greenland (Johnsen et al., 1972). Stable isotope records from Greenland cover time intervals up to 250 000 years and from Antarctica - up to 750 000 years. Correlation between $\delta^{18}O$ ice core reconstructions and instrumental series reaches 0.3–0.4 over annual time scale.

The isotopic composition of wood cellulose is another area of paleoisotope temperature reconstructions. Biological and physical processes determine stable isotope $^{13}C/^{12}C$ ratios in organic matter during photosynthetic uptake of CO_2 from the air (Farquhar et al., 1982) while the ratio $^{18}O/^{16}O$ is determined by isotopic composition of source water to the tree, evaporation effects in tree leaf and biochemical steps during cellulose synthesis (Roden et al., 2000). ^{13}C abundance in tree-ring cellulose is expressed as $\delta^{13}C$ values relative to the Vienna PDB standard ($[^{13}C]/[^{12}C]= 0.0112372$). Strong response of $\delta^{13}C$ in northern Finland (Kessi) to midsummer temperature was found by Hilasvuori et al., (2009). Coefficients of correlation between the $\delta^{13}C$ record from Kessi and temperature, measured at the weather stations, is 0.68 at annual time scale and 0.81 at decadal time scale. Thus the stable carbon tree-ring series with annual resolution are valuable for further paleoclimatic research. It should be noted that recent studies have shown that $\delta^{13}C$ record is not only a pure temperature proxy but sunshine too (Gagen et al., 2011).

Ocean coral skeletal rings, or bands, also impart paleoclimatological information. Cooler temperatures as well as denser water salinity tend to cause coral to use heavier isotopes in its structure. Deep ocean sediments have been examined to get information about the conditions during the past 1 000 000 years

3.3 Other sources of paleoclimatic information

Borehole temperature measurement, paleobotanic data, historical data and melt layer thickness are other proxies used by paleoclimatology. Subsurface terrestrial borehole temperature profiles can be used to obtain an estimate of ground surface temperature changes back in time. An advantage of the borehole data over those from the majority of other climate proxies (tree rings, corals, ice cores, and historical documentary records), is that they do not require calibration against independent surface temperature data. As opposed to tree-rings, borehole temperatures are only sensitive to climate variations at multi-decadal or longer time scales due to the attenuation by the heat diffusion process. Borehole data have been recently used to characterize Northern Hemisphere continental temperature for the past 500 years (Beltrami., 2002; Mann et al., 2003).

Pollen grains which are washed or blown into lakes and peatlands can accumulate in sediments. Plants produce pollen in large quantities and it is extremely resistant to decay. It is possible to identify a plant species from its pollen grain. The identified plant community of the area at the relative time from that sediment layer will provide information about the climatic condition. Pollen-based records are considered sensitive to multi-decadal variability. Pollen analysis has been used to derive quantitative information about the climate of the past 200 000 years.

Contemporary written historical records - diaries, annals - contain information on a variety of natural phenomena: freeze dates, harvest amounts, flowering dates, extreme droughts, hurricanes etc. These data are often highly subjective and unhomogeneous. Thus their analysis is a complicated task. However, promising methodology has been developed to produce centennial time series on past climate extremes (Dobrovolny et al., 2010).

Ice core melt layers store information about summer temperature. The data on melt layer thickness might not reflect the full range of the temperature variability - very cold summers could cause no melt while during very warm summers the whole layer will melt. Thus, coefficient of correlation between melt layer proxies and instrumental temperature often is less than 0.2.

Since different paleoindicators reflects actual temperature changes by different ways the *multiproxies* - the time series, which generalize proxy sets of various types - are often used by paleoclimatology. For example, M. Mann and his colleagues obtained their famous reconstruction using 12 proxy indicators - 7 tree-ring width based, 2 tree-ring density based, 2 based on $\delta^{18}O$ in ice core, 1 based on ice accumulation rate (Mann et al., 1999). McCarroll et al. (2003) used multiproxy analysis to several annually-based series from the same trees at treeline. Moberg et al. (2005) combined proxies with low time resolution (pollen, sediments, borehole) data with high-resolution (tree-ring data) using a wavelet technique. Several recent temperature reconstructions provide information about large-scale climate variability over the past 1000 years (Fig. 5). These paleoclimatic proxies include the following temperature records: the multiproxy for the Northern Hemisphere by Mann et al. (1999), the multiproxy for the Northern Hemisphere by Jones et al. (1998), the multiproxy for the Northern Hemisphere by Crowley and Lowery (2000), the extratropical Northern Hemisphere tree-ring proxy by Esper et al. (2002), the tree-ring proxy for the northern part of the Northern Hemisphere by Briffa (2000), the multiproxy for the extratropical Northern Hemisphere produced by Moberg et al. (2005), the multiproxy for the Northern Hemisphere by Loehle (2007) which does not include any dendroclimatological information. Loehle (2007) used the data on pollen, stable ^{18}O in ice cores and sediments, speleotemperature, Mg/Ca ratio and estimations based on diatoms and planktonic foraminifera. Ogurtsov and Lindholm (2006) showed that millennial climatic records have strong discrepancies in the low frequency domain and, thus, demonstrate different histories of temperature changes during the last 1000 years. They can be divided into three clusters:

a. The reconstructions of Mann et al. (1999), Jones et al. (1998) and Crowley and Lowery (2000) (Figure 5A,B,C) show an obvious linear decrease of mean temperature until the middle of the 19th century and a sharp rise thereafter - the so-called "hockey-stick" form. These records show unambiguously that Earth in the 20th century is warmer than at any time during the last millennium.
b. The reconstructions of Briffa (2000) and Esper et al. (2002) (Figure 5D,E) do not show a similar linear trend. Instead of this, multi-centennial long-term changes dominate, and the recent warming does not seem to be anomalous. To a degree they represent natural climate cycles.
c. In reconstructions of Moberg et al. (2005) and Loehle (2007) (Figure 5F,G) millennial-scale cycles prevail and the twentieth century is not extremely warm.

In the work of Ogurtsov et al. (2011) it was shown that in spite of discrepancy between different proxies they all testify that that in the extratropical part of the Northern Hemisphere the time interval 1988-2008 was likely the warmest two decades throughout the last 1000 years.

A major problem in paleoclimatic studies is short interval of calibration due to limited instrumental data - usually no more than 90-100 years. It is not enough to make confident judgment about the quality of reconstructed long-term climate changes (periods of hundreds of years or more). Further the time of the strongest climate signal, e.g. temperature, during the instrumental period may vary markedly within growing season in the 100-year calibration period (Tuovinen et al., 2009), thus affecting past reconstructions as well. Also divergence, i.e. e.g. temperature-proxy relation becomes opposite by periods (e.g. Seo et al., 2011), increases uncertainty in the reconstructions of the past climate beyond the instrumental period. High-resolution decline in the time series may be interpreted as a short-term dry or cold period although low indices in the series were caused by sudden loss of needles by pests or pathogens (Ferretti et al., 2002).Therefore the current ability of paleoclimatiology to reconstruct precisely the long-term - potentially the most powerful - climate variations is disputable.

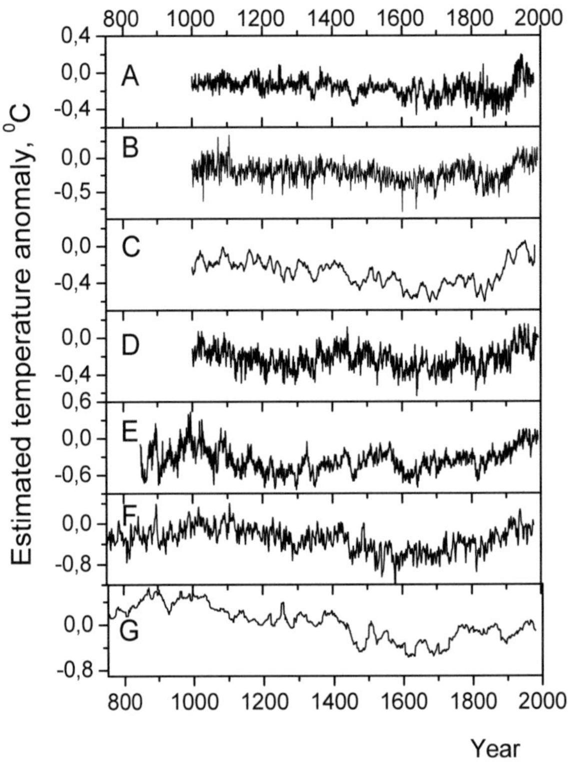

Fig. 5. Millennial-scale temperature proxies, obtained by: A – Mann et al. (1999), B – Jones et al. (1998), C – Crowley and Lowery (2000), D – Briffa (2000), E – Esper et al. (2002), F – Moberg et al. (2005), G – Loehle (2007). All the data sets were adjusted to the mean extratropical instrumental temperature for 1880-1980.

4. Solar irradiance and terrestrial climate

Solar radiation is practically the only source of energy for the terrestrial atmosphere. Therefore the most evident direct effect of solar variability on climate is its influence on the Earth's radiative balance through variations in luminosity. It has been established that TSI follows the general 11-year cycle of solar activity (see Fig. 1E). The amplitude of this variation, however, is small – ca. 2 Wm^{-2}. It is not enough to effectively influence climatic processes, particularly if taken into account that short-term oscillations of energy input are substantially attenuated by the thermal inertia of oceans, which has large heat capacity and integrates variations in heat input. The attenuation of long-term (multidecadal and longer) variations is weaker. However it is unknown whether TSI has such slow variability. Direct satellite measurements over 2-3 cycles are not long enough for any decisive conclusion. Nevertheless many prolonged TSI reconstructions have been obtained by different researchers using sunspot numbers and radioisotopes as proxies (Fig. 6).

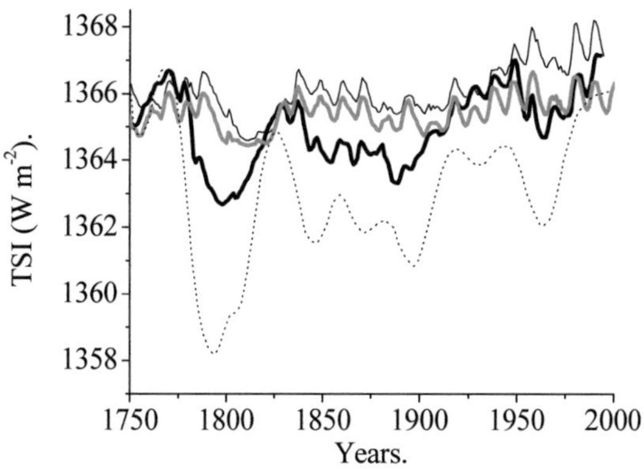

Fig. 6. Reconstructions past variations in the total solar irradiance obtained by: Hoyt and Schatten (1993) - thick black line; Lean et al. (1995) - thin black line; Beer et al. (2000) - dotted line; Mordvinov et al. (2004) - grey line.

During the 20th century, the increase in solar irradiance could have been 1-3 W m^{-2} or more (Fig.6). That is enough to provide the radiative forcing of 0.18-0.52 W m^{-2}. The IPCC estimates a smaller value – according to IPCC (2007) changes in solar irradiance since 1750 is expected to cause a radiative forcing of 0.06–0.30 W m^{-2}. Let us evaluate the probable climatic response to this forcing. For a rough quantitative estimation we can multiply radiative forcing by climatic sensitivity λ_c. Estimations obtained using different methods, including analysis of instrumental temperature data, paleodata, the data of the satellite Earth Radiation Budget Experiment (ERBE) etc., are listed in Table 1.

Author	Way of estimation	λ_c (^{0}C W^{-1} m^2)
Lindzen and Giannitsis (1998)	Climatic response to volcanic explosions (Krakatau, Katmai, Pinatubo)	0.07
Idso (1998)	8 natural phenomena (equator to pole temperature gradient, faint early Sun paradox, etc.)	0.1
Lindzen and Choi (2009)	1985 to 1996 ERBE data, sea surface temperature	0.14
Lindzen et al. (2001)	Upper-level cloudiness and sea-surface temperature data from the tropical Pacific	0.17–0.43
Schwartz (2008)	Global surface temperature	0.25–0.79
Chylek et al. (2007)	Global surface temperature, CO_2, aerosol optical depth after 1985	0.29–0.48
Chylek end Lohmann (2008)	Antarctic paleodata (temperature, CO_2, CH_4, dust) of LGM to Holocene transition	0.36–0.68
IPCC (2007)	Various ways of estimation	0.53–1.23
Andronova and Schlesinger (2001)	Global mean and hemispheric difference in surface air temperature 1856 - 1997	0.27–2.54
Frame et al. (2005)	Global surface temperature	0.32–3.14
Forster and Gregory (2006)	1985 to 1996 ERBE data 60°N to 60°S, global surface temperature	0.32–3.78

Table 1. Estimations of climate sensitivity.

Available assessments of λ_c differ by more than an order of magnitude (see Table 1). This is caused by the complexity of climatic system which has a lot of feedbacks. Many of these feedbacks are currently not well known. If we consider the estimations of IPCC (2007) we obtain that the increase of solar brightness in the 20th century could have resulted from corresponding global warming by 0.10-0.64^{0} C. The difference between the lower and upper limits of the solar induced increase in the global temperature reaches a factor six. If we use the sensitivity evaluations not only of IPCC but also of other authors the difference will be even more. Such uncertainty does not allow us to draw any decisive conclusions about the role of the Sun in global warming. It may be significant as well as negligible. This result is in agreement with the IPCC (2007) considering current level of scientific understanding of TSI-climate link as low.

Variation in ultra-violet (UV) solar radiation, which effectively influences ozone layer, is another way to provide a link between solar luminosity and climate. The amplitude of the 11-year variation in UV radiation reaches several percent (ca. 6% at 20 nm). Ozone is the main gas involved in the radiative heating of the stratosphere. Solar-induced changes in ozone can therefore affect the radiative balance of the stratosphere with indirect effects on the troposphere and global circulation pattern. Possible reaction of climatic system to the UV solar variability was analyzed by Haigh (1996), Shindell et al. (1999), Gray et al. (2010).

5. Cosmic ray flux and terrestrial climate

The Sun can influence terrestrial weather and climate not only directly through variations of TSI but indirectly - via modulation of corpuscular radiation entering atmosphere. As we have noted above, the idea that cosmic ray flux could directly influence the weather was proposed by Ney (1959) and developed by Dickinson (1975). However the total energy input of the solar-modulated particles is very small as compared to the energy of atmospheric processes - see Table 2 (the data were taken from Vitinsky et al., 1986; Pudovkin, 1996 and Borisenkov, 1977).

Type of energy	Power (erg/day)
Energy brought in the Earth's atmosphere by TSI	1.6×10^{29}
Energy brought in the Earth's magnetosphere by solar wind	up to 3×10^{23}
Energy brought in the Earth's magnetosphere by the fast solar wind during magnetic storms	up to 4×10^{24}
Energy brought in the Earth's atmosphere by the flux of GCR with E>0.1 GeV	$\sim10^{22}$
Energy brought in the Earth's atmosphere by the water vapor (latent heat)	4×10^{28}
Energy release by thunder clouds	up to 10^{27}
Power of the cyclone (anticyclone)	up to 10^{25}

Table 2. Characteristic energy of the terrestrial manifestations of solar activity and atmospheric processes.

It is evident from Table 2 that the energy coming into terrestrial magnetosphere and the atmosphere is a few orders of magnitude less than the energy of meteorological processes. Energy input from the very large solar flare of August 1972 is also small - ca. 10^{24} erg. For this reason a linear relationship between cosmic rays and the Earth's climate seems very unlikely. However, the effect of cosmic radiation on the atmosphere could be very nonlinear. For example, minor changes in the physical and chemical parameters of the atmosphere, which does not require a lot of energy, could appreciably change its optical properties, disturb the energy balance and stimulate powerful atmospheric processes. Indeed, Pudovkin and Veretenenko (1992) have studied changes in the atmospheric transparency following the geomagnetic disturbances, caused by the fluxes of solar and galactic cosmic particles. They showed that corresponding changes in energy balance of the lower atmosphere can reach 1.5×10^{26} erg/day. Starkov and Roldugin (1995) have obtained even a larger value - ca. 10^{27} erg/day. These estimations testify that cosmic radiation in fact can influence terrestrial weather. Since the intensity of cosmic rays is driven by solar activity the proposed mechanism of indirect solar-climate link appears quite plausible. A lot of experimental evidences of the reality of a link between the fluxes of cosmic particles and the physical state of the Earth's atmosphere have been obtained (Table 3).

In Table 3 CR-BR↑↓ indicates negative correlation between cosmic ray flux and background radiation i.e. positive correlation between cosmic ray intensity and cloudiness or/and aerosol content. This effect is considered as a manifestation of the *optical* mechanism of the solar-climate link. It is believed that the optical mechanism is caused by the influence of atmospheric ionization on production of new aerosol particles and cloud microphysics. Abbreviation CR-AC indicates a link between cosmic ray flux and atmospheric circulation.

Despite the evidence for the reality of a link between cosmic rays and weather and climate the time-period of the experimental studies is rather short - usually no more than 30- 40 years. Even within this short time interval results are equivocal - e.g. some powerful flares were accompanied by increase of aerosol content in the atmosphere while some other flares did not show the effect. Extending data on low cloudiness (International Satellite Cloud Climatology Project - ISCCP) towards 2007 substantially reduces correlation between GCR and low-level clouds (see Fig. 15 in Gray et al., 2010). Moreover Roldugin and Tinsley (2004) reported that during 1978–1989 atmospheric transparency above 45 stations, situated at former USSR territory, decreased (not increased) during Forbush decreases, when sulfate aerosol loading to the stratosphere was high. However, Calgovic et al. (2010) find no response of global cloud cover to Forbush decreases at any altitude and latitude. Because of the lack of experimental data and some controversy in experimental results more information is needed before definite conclusions may be drawn. We can use the paleodata on nitrate concentration [NO_3^-] and the conductivity of an ice core extracted by specialists from the University of Kansas in (central Greenland, GISP2 H core, 72°N, 38°W, height 3230 m) for this purpose. Both glaciochemical series cover the time period of 1576–1991 and have extremely high time resolution (approximately 20 samples per year) (Dreschhoff and Zeller, 1994). As we have noted above nitrate concentration in ice could serve as a proxy of atmosphere's ionization. The conductivity of ice is connected with its acidity and, hence, reflects concentration of the sulfate aerosol in the atmosphere. Some [NO_3^-] peaks coincide in time with conductivity bursts (Fig. 7). Ogurtsov (2011) analyzed such events in 1789, 1859, 1895, 1896, and 1908 and showed that the abrupt increase in the concentration of sulfate aerosols in the stratosphere due to additional ionization as a result from precipitation of solar cosmic ray energetic particles is one of the most probable factors that cause simultaneous origination of powerful peaks in the ice conductivity and nitrate concentration. Dreschhoff and Zeller (1994) have related the phenomenon to the direct effect of ionization on the formation of nacreous or polar stratospheric clouds (PSCs), which are largely composed of nitric acid (HNO_3) droplets. Thus, coincidences of peaks in both

Authors	Time inter-val of obser-vation	Type of the data	The revealed effect	Geographic coverage	Time scale of the effect	Type of effect
Pudovkin and Veretenenko (1992)	1961-1984	A	Increase of direct background solar radiation during FDs	USSR, $=60^{0}-67^{0}$ N	days	CR-BR
Roldugin and Vashenyuk (1994)	1972, 1976, 1979	A	Decrease of atmospheric transparency 1-3 days after SCR events	Murmansk, Arkhangelsk, Leningrad	days	CR-BR
Todd and Kniveton, (2004)	1983-2000	S	Reduction of high-level cloudiness after FDs	Antarctica	days	CR-BR
Tinsley and Deen (1991)	1953-1985	M	Link between FDs and winter cyclone intensity	Northern Hemisphere	days	CR-AC
Veretenenko and Pudovkin (1993)	1970-1986	M	Intensification of atmospheric circulation after SCR events and reduction after FDs	$=45^{0}-65^{0}$ N	days	CR-AC
Shumilov et al. (1996)	Febru-ary 1983	L	Increase of stratospheric aerosol concentration at 15-25 km after GLE event	Verkhne-tulomsk	days	CR-BR
Marichev et al. (2004)	March 1988, March 1989	L	Increase of stratospheric aerosol concentration at 40-45 km after geomagnetic disturbances	Tomsk	days	CR-BR
Svensmark et al. (2009)	1987-2007	S	Reduction in liquid cloudi-ness and concentration of fine aerosol particles after FDs	Globe	days	CR-BR
Laken et al, (2010)	1986-2006	S	Reduction of a total cloudiness after FDs	$=30^{0}-60^{0}$ N $=30^{0}-60^{0}$ S	days	CR-BR
Kondratyev and Nikolsky (1983)	1962-1970	R	Decrease of atmospheric transparency above 25 km when Wolf number rises from 50-70 to 170	Leningrad	10 years	CR-BR
ISCCP (Marsh and Svensmark, 2000; Palle et al., 2004)	1984-2004	S	Negative correlation between low-level cloudiness and GCR intensity	Globe	10 years	CR-BR
Lu et al. (2007)	1958-2004	M,R	Relationship between solar irradiance and geomagnetic activity and the atmospheric temperature.	Globe	Months-decades	CR-BR
Veretenenko et al. (2007)	~1880-1970	M	Correlation between surface pressure and Wolf number	North Atlantic $=20^{0}-40^{0}$ N $=45^{0}-65^{0}$ S	10-100 years	CR-AC

Table 3. Experimental evidences for a link between solar modulated cosmic radiation and physical state of the Earth's atmosphere. Abbreviation A means data of actinometrical stations, M - data of meteorological stations, L - data of lidar observations, R- radiosonde data, S - satellite data.

studied glaciochemical series are manifestations of the effect that has been experimentally registered with lidar and satellite equipment for the last 25 years. This proves that the relationship between the aerosol concentration and the ionization rate in the stratosphere is real and makes it possible to extrend the interval where this connection exists for more than 200 years. The result provides new evidence of the reality of optical mechanism of the Sun's influence on weather.

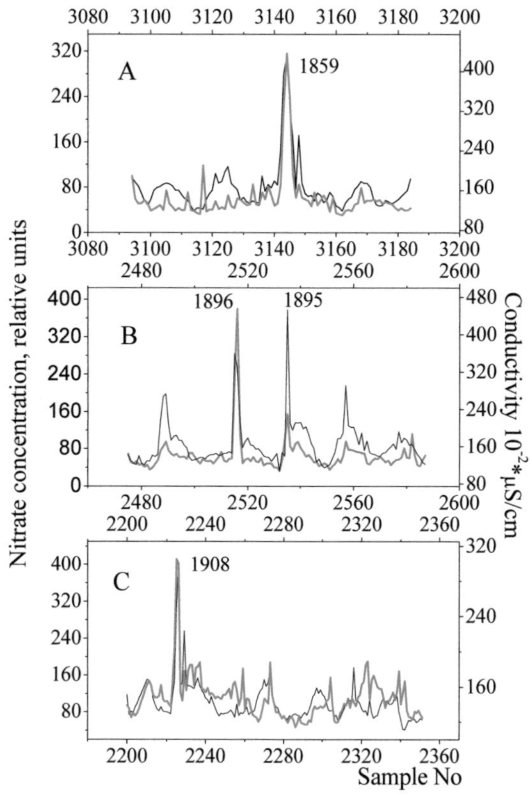

Fig. 7. Concentration of ions (thin black line) and conductivity (thick grey line): (a) in 1859 (the Carrington optical flare), (b) at the end of the 19th century, and (c) in 1908 (the Tunguska event). 100 relative units = 90.3 ng/g NO_3^-.

Optical mechanism could also provide solar-climate relationship over longer time scale. The effect of unphysical time lag between long-term variation in global temperature and Wolf number can also be explained in the framework of this mechanism. It has been established that long-term (T>30 yr) variation in global temperature resemble the variation in sunspot number since the mid-19th century which might be attributed to a solar-climatic connection (see Fig. 8). However, smoothed temperature outstrips solar activity that violates the cause-effect relationship. Friis-Christensen and Lassen (1991) have shown that this phenomenon could be explained by using solar cycle length (SCL) as an indicator of solar activity instead of sunspot numbers - SCL lead sunspot number by 1- 3 cycles. However, physical processes responsible for this link have not been established so far, which calls for the search for alternative explanations.

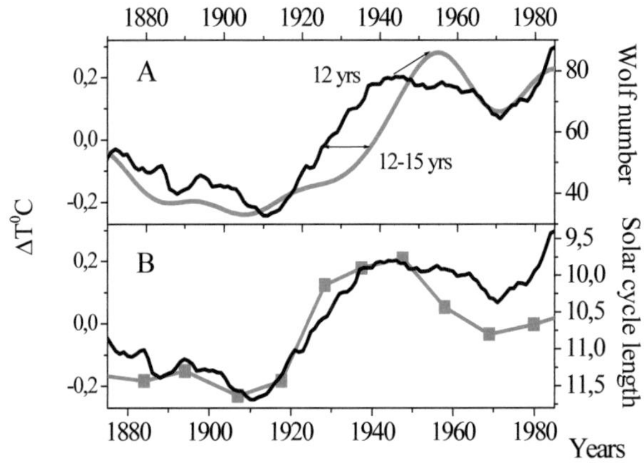

Fig. 8. A - gray curve - Wolf numbers smoothed by the Furier-filter (frequencies above 0.04 years^{-1} are suppressed). Black curve - annual temperature of Northern hemisphere (Jones et al., 2001) smoothed by 15 years. B - gray curve - the length of a solar cycle smoothed on three points. Black curve - the smoothed annual temperature of Northern hemisphere.

Ogurtsov et al. (2004) have revealed a negative correlation between century-type variations in NO_3^- concentration in Greenland ice and sunspot number. This study found that the Gleissberg cycle in nitrate leads the corresponding variation in sunspot number by 15–18 years during the last 3 centuries. Since [NO_3^-] in polar ice is connected with the stratosphere's ionization it is reasonable to assume that the century-long cycle in stratospheric ionization advances the corresponding cycle in solar activity. Experimental estimation of the aerosol optical depth of the atmosphere, performed by Bryson and Goodman (1980) using data of 42 actinometrical stations situated within 25^0–65^0 N belt, showed the presence of distinct multidecadal variation (Fig. 9). Long-term changes of atmospheric transparency are generally the result of the corresponding change in the optical depth of the stratosphere. The century-type periodicity in the stratospheric aerosol transparency correlates negatively with the corresponding cycle in sunspots and lead it by 20-25 yrs (Figure 9B). Aerosol in the stratosphere is basically sulfate with a composition of about 75% H_2SO_4 and 25% H_2O. Volcanoes are the main source of SO_2 in stratosphere, from which H_2SO_4 is generated. Long-term variation in volcanic activity lags behind aerosol transparency instead of advancing it (Fig. 9C). Volcanic SO_2 loading into the stratosphere estimated by Kondratyev (1999) does not correlate well with the aerosol optical depth (Fig. 9D). This means that long-term variation of the sulfate aerosol abundance in the stratosphere is influenced not only by volcanic SO_2 input but also by one additional factor. Ogurtsov (2007) has hypothesized that this factor might be a century-long cycle in stratospheric ionization, which, in turn, is caused by superposition of the century-scale variations in GCR and SCR intensity. I.e. long-term variation of aerosol optical depth of stratosphere is driven by corresponding change in solar-cosmic corpuscular radiation, which produces century-scale cycle in stratospheric ionization advancing the Gleissberg periodicity in sunspot number. The estimation by Ogurtsov (2007) using energy-balance model showed that the observed variation in the atmosphere's transparency during the 20th century is enough to provide the global temperature change. Thus, according to Ogurtsov (2007) the positive time lag between long-term variations in surface temperature and Wolf number could not be linked to the change in solar cycle length but to the variation in stratospheric optical depth. The advancing development of century-long cyclicity in temperature is a result of the following chain of physical processes: solar activity → fluxes of ionizing particles (SCR and GCR) → ionization → number of ultra-fine particles in stratosphere → concentration of stratospheric aerosol → atmospheric transparency → background solar radiation → surface temperature. Thus the effect of positive time shift between century-long variations in temperature and sunspot number could be explained in the framework of optical mechanism of solar-climate link. The most serious problem of the proposed mechanism is the lack of knowledge about the physical processes responsible for a link between ionization and the number of condensation nuclei (CN) and cloud condensation nuclei (CCN). Ion-induced particle formation is a plausible mechanism to provide this link. New aerosols initially appear as small (<2 nm) clusters containing a few molecules. These clusters either may grow by further condensation of nearby gas molecules or evaporate. Above a certain critical size, the cluster is thermodynamically stable and its probability to grow by further condensation is larger than the probability to break up by evaporation. Atmospheric ions enhance the process of clustering of condensable vapor. The presence of a charge stabilizes the embryonic cluster through Coulomb attraction, and reduces the critical size (Yu and Turco, 2000, 2001). The term *ion-induced nucleation* (IIN) describes the formation of new ultra-fine particles from the gas phase, in which ions take a part. The theory of this process was developed in the works of Arnold (1982, 2006), Kazil and Lovejoy (2004) and Yu (2006). It has been shown that although INN is connected directly with ionization, it depends also on the physical parameters of the atmosphere, particularly on the concentration of sulfuric acid vapor [H_2SO_4] and temperature. High values of [H_2SO_4] (about 10^7 cm^{-3} or more) and low (<-50^0 C) temperature are necessary for effective ion-induced generation of ultra-fine aerosol particles. Therefore solar activity affects the atmosphere in the framework of INN mechanism not directly but its influence is mediated by a few internal terrestrial phenomena - volcanic activity, meteorology etc. Such a complex solar-atmosphere link might be the cause of instability of solar-climatic correlations.

The charge also accelerates the early growth process, due to an enhanced collision probability. As a result the charge appreciably helps the sub-critical embryos to survive and grow to CN and CCN sizes. Thus, despite the IIN mechanism is believed to work mainly under clear sky conditions, it can also communicate solar activity to cloudiness. Experimental evidence for IIN have been obtained by Eichkorn et al. (2002); Lee et al. (2003) and Svensmark et al. (2007). It was shown that changes in the global atmospheric circuit can influence cloud microphysics (Tinsley, 2000). The global circuit causes a vertical current density in fair (non-thunderstorm) weather regions. This ionosphere-surface fair weather current density passes through clouds causing local droplet and aerosol charging at their boundaries, where sharp gradients in air conductivity can occur. Modulation of the global circuit by solar-induced changes in atmospheric ionization (both GCR and SCR effect) provides an alternative plausible route by which solar changes can be transmitted to the lower atmosphere (Tinsley, 2000, 2008). Three potential near-cloud mechanisms of the influence of the conduction current density on clouds have been proposed:

a. *electroscavenging* is a result of electrically-enhanced efficiency of the collision between charged particles with liquid droplets (e.g. Tinsley et al., 2008).
b. *Electrofreezing* is connected to the possible influence of electric properties of aerosol on the rate of freezing of thermodynamically unstable super-cooled water droplets (Tinsley and Dean, 1991).

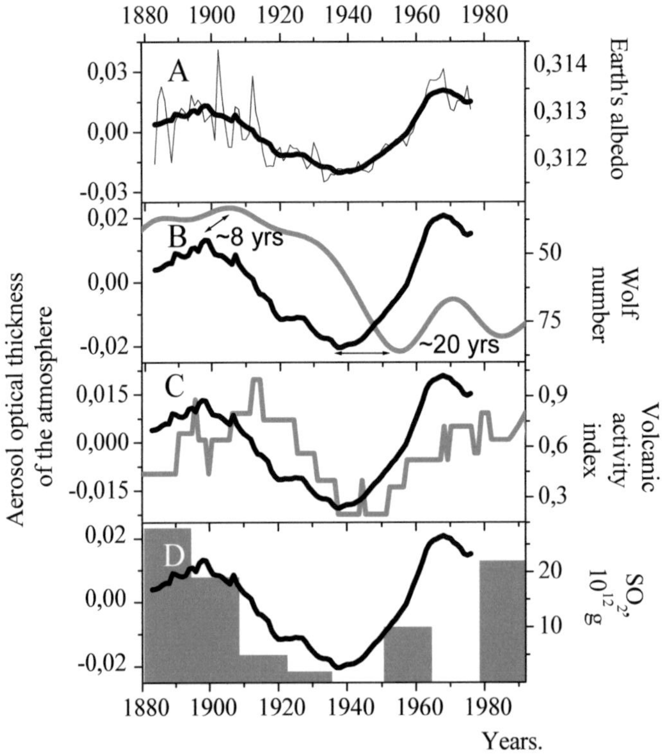

Fig. 9. A – aerosol optical thickness of a middle latitude (25⁰-45⁰) stratosphere after Bryson and Goodman (1980) annually interpolated. Thin curve - raw data, thick curve - data smoothed by 11 years; B – gray curve - Wolf number smoothed by the Fourier filter (frequencies above 0.04 years^{-1} are suppressed), black curve - aerosol optical thickness of a stratosphere, smoothed by 11 years; C – gray curve - volcanic explosive index after Briffa et al. (1998) smoothed by 25 years. Black curve – smoothed aerosol transparency; D – gray columns – SO_2 loading to stratosphere after Kondratyev (1999).

c. *Electrocoalescence* is connected to the possible increase in coalescence between charged droplets, which could influence droplet size or number (Harrison and Ambaum 2009).

As we have noted above, all the mechanisms work at the boundaries of super-cooled and liquid clouds. Direct experimental evidence for a link between atmospheric electricity and cloud microphysics were obtained by Harrison and Ambaum (2009) as well as Nicoll and Harrison (2009). It should be considered also that the global circuit is dependent on atmospheric column resistance, which, in turn is influenced not only by ionization but also by concentration of sulfate aerosol in the stratosphere, concentration of cosmic dust in upper atmosphere and, probably other factors. The contribution of these factors might obscure solar impact on the vertical current density.

A new mechanism for the connection between solar-modulated corpuscular radiation and physical processes in the lower atmosphere was proposed by Avakyan and Voronin (2010).

They assumed that microwave radiation of the Earth ionosphere during solar flares and geomagnetic storms might influence the condensation processes in the troposphere and thus influence the weather.
Our understanding of possible physical mechanisms of influence of the flux of solar-cosmic corpuscles on the atmosphere of the Earth and influence of solar and geomagnetic activity on weather and climate has considerably improved during the last 2-3 decades.

6. Conclusion and prospects for further research

The last few decades were marked by considerable successes in helioclimatology. Both satellite and ground-based observations have brought a lot of new evidence for a link between solar activity and the phenomena of the lower atmosphere. The progress in theory, experiment and modeling has also significantly increased our knowledge of the Sun and solar-terrestrial connections. Nevertheless, the absolutely conclusive proof of the reality of the solar-climate link is still missing. The lack of facts and understanding about the connecting processes at work is the main cause of this shortcoming. Data obtained by experimental observation are quite precise but rather short in time scale. Paleoproxies are much longer but their uncertainty generally increases as a function of time from the present. Thus, in solar-climatic research we face a kind of "uncertainty principle":

$$\frac{\Delta X}{\Delta T} \approx \text{const}, \tag{7}$$

where ΔX is the uncertainty of the data, and ΔT is the length of the X time series.
Moreover, it has been shown that the Sun-climate connection, even if it actually exists, may be realized by an indirect and nonlinear way. As a result the search for a link between solar activity and weather and climate has turned out to be a more difficult task than was previously anticipated. Quite possibly further future research in helioclimatology will follow the guidelines envisaged below:

1. Further gathering and accumulation of the information together with its subsequent systematization. This concerns both instrumental monitoring of solar-cosmic and geophysical parameters and work on constructing new paleoreconstructions of solar activity and climate. The development of methods of paleoastrophysics and paleoclimatology is an integral part of this work.
2. The improvement of the methods of statistical analysis, particularly, methods aimed at the search and detection of nonlinear interrelations between the different time series.
3. Further laboratory research of the physical processes which possibly provide a link between solar activity and the low atmosphere.
4. Increasing our understanding about the climatic system and improvements of methods of climatic modeling.

7. Acknowledgment

M. G. Ogurtsov expresses his thanks to the exchange program between the Russian and Finnish Academies (project No. 16), to the program of the Saint-Petersburg Scientific Center of RAS for 2011, and to RFBR grants No. 09-02-00083, 10-05-00129, 11-02-00755 for financial support. R. Jalkanen and M. Lindholm want to thank Academy of Finland for its financial support (SA 138937).

8. References

Andronova, N.G. & Schlesinger, M.E. (2000). Causes of Global Temperature Changes During the 19th and 20th Centuries. *Geophysical Research Letters*, Vol. 27, No.14, pp. 2137-2140, ISSN 0094-8276

Arnold, F. (1982). Ion Nucleation - a Potential Source for Stratospheric Aerosols. *Nature*, Vol. 299, pp. 134-137, ISSN 0028-0836

Arnold, F. (2006). Atmospheric Aerosol and Cloud Condensation Nuclei Formation: A Possible Influence of Cosmic Rays. *Space Science Reviews*, No.1-4, pp. 169-186, ISSN 0038-6308

Avakyan, S. V. & Voronin, N. A. (2010). Radio-Optical and Optical Mechanisms of the Influence of Space Factors on Global Climate Warming. *Journal of Optical Technology*, Vol.77, No. 2, pp. 150-152 , ISSN 1070-9762

Bard, E.; Raisbeck, G.; Yiou, F. & Jouzel, J. (2000). Solar Irradiance During the Last 1200 Years Based on Cosmogenic Nuclides. *Tellus B*, Vol. 52, No.3, pp. 985-992, ISSN 1600-0889

Bazilevskaya, G.A.; Usoskin, I.G.; Fluckiger, E.O.; Harrison, R.G.; Desorgher, L.; Butikofer, R.; Krainev, M.B.; Makhmutov, V.S.; Stozhkov, Y.I.; Svirzhevskaya, A.K.; Svirzhevsky, N.S. & Kovaltsov, G.A. (2008). Cosmic Ray Induced Ion Production in the Atmosphere. *Space Science Reviews*, DOI 10.1007/s11214-008-9339-y, ISSN 0038-6308

Beer, J.; Mende, W. & Stellmacher, R. (2000). The Role of the Sun in Climate Forcing, *Quaternary Science Reviews*, Vol.19, pp. 403-415, ISSN 0277-3791

Beltrami, H. (2002). Climate from Borehole Data: Energy Fluxes and Temperatures Since 1500. *Geophysical Research Letters*, Vol.29, No.23, pp. 2111, doi:10.1029/2002GL015702, ISSN 0094-8276

Bond, G.; Kromer, B.; Beer, J.; Muscheler, R.; Evans, M.; Showers, W.; Hoffmann, S.; Lotti-Bond, R.; Hajdas, I. & Bonani, G. (2001).Persistent Solar Influence on North Atlantic Climate During the Holocene. *Science*, Vol.294, No.5549, pp. 2130 - 2136, ISSN 1095-9203.

Borisenkov, E.P. (1977). Development of Fuel and Energy Base and its Influence on Weather and Climate. *Meteorology and Hydrology*, No.2, pp.3-14, ISSN 0130-2906, (in Russian)

Briffa, K.R.; Jones, P.D.; Schweingruber, F.H. & Osborn, T.J. (1998). Influence of Volcanic Eruptions on Northern Hemisphere Summer Temperature Over the Past 600 Years. *Nature*, Vol. 393, No.6684, pp. 450-455, ISSN 0028-0836

Briffa, K.R. (2000). Annual Climate Variability in the Holocene: Interpreting the Message of Ancient Trees. *Quaternary Science Reviews*, Vol. 19, pp. 87-105, ISSN 0277-3791

Briffa, K.R. & Osborn, T.J. (2002). Blowing Hot and Cold. *Science*, No.5563 , Vol.295, pp. 2227-2228, ISSN 1095-9203

Bryson, R.A. & Goodman, B.M. (1980). Volcanic Activity and Climatic Changes. *Science*, Vol.207, No.7 , pp. 1041-1044, ISSN 1095-9203

Calogovic, J.; Albert, C.; Arnold F.; Beer J.; Desorgher L. & Flueckiger, E. O. (2010), Sudden cosmic ray decreases: No change of global cloud cover. *Geophysical Research Letters*, Vol.37, L03802, doi:10.1029/2009GL041327, ISSN 0094-8276

Chizhevsky, A.L. (1973). *Terrestrial echo of solar storms*, Mysl', Moscow (in Russian)

Chylek, P.; Lohmann U.; Dubey M.; Mishchenko M.; Kahn, R. & Ohmura, A. (2007). Limits on Climate Sensitivity Derived From Recent Satellite and Surface Observations.

Journal of Geophysical Research, Vol.112, D24S04, doi:10.1029/2007JD008740, ISSN 0094-8276

Chylek, P. and Lohmann, U. (2008). Aerosol Radiative Forcing and Climate Sensitivity Deduced From the Last Glacial Maximum to Holocene Transition, *Geophysical Research Letters*, Vol.35, L23703, doi:10.1029/2008GL033888, ISSN 0094-8276

Cook, E.R. & Kairiukstis, L.A. (1989). *Methods of Dendrochronology: Applications in the Environmental Sciences,* Kluwer Academic Publishers, Dordrecht

Crowley, T.J. & Lowery, T.S. (2000). How Warm Was the Medieval Warm Period? *Ambio*, Vol. 29, pp. 51-54, ISSN 0044-7447

Dansgaard, W. (1954). The O18-abundance in Fresh Water. *Geochimica et Cosmochimica Acta*, Vol. 6, No. 5-6, pp. 241-260, ISSN 0016-7037

De Jager, C. (2005). Solar Forcing of Climate: Solar Variability. *Space Science Reviews*, Vol.120, pp.197–241, ISSN 0038-6308

Dergachev, V.A.. (1994). Radiocarbon chronometer. *Priroda*, No. 1, pp. 3-15, ISSN (in Russian).

Dickinson, E. (1975). Solar Variability and the Lower Atmosphere. *Bulletin of American Meteorological Society*, Vol.56, pp.1240-1248, ISSN 1520-0477

Dreschhoff, G.A.M.; Zeller, E.J. & Parker, B.C. (1983). Past Solar Activity Variation Reflected in Nitrate Concentration in Antarctic Ice. In: *Weather and climate response to solar variations*, B.M. Mc Cormac, (Ed.), 225-236, Colorado Associated University Press, Boulder

Dreschoff, G.A.M. & Zeller, E.J. (1994). 415-year Greenland Ice Core Record of Solar Proton Events Dated by Volcanic Eruptive Episodes. In: *TER-QUA Symposium Series 2*, D. Wakeffield, (Ed.), 1-24, Nebraska Academy of Sciences, Lincoln, Nebraska

Dreschoff, G.A.M. & Zeller, E.J. (1998). Ultra-high Resolution Nitrate in Polar Ice as Indicator of Past Solar Activity. *Solar Physics*, Vol. 177, pp. 365-374, ISSN 0038-0938

Douglass, A. E. (1914). A Method of Estimatig Rainfall by the Growth of Trees. In,: *The climatic factor*, E. Huntington, (Ed.), 101-122, Carnegie Institute Washington Publications.

Dobrovolny, P.; Moberg, A.; Brazdil, R.; Pfister, C.; Glaser, R.; Wilson, R.; van Engelen, A.; Limanowka, D.; Kiss, A.; Halickova, M.; Mackova, J.; Luterbacher, J. & Bohm, R. (2010). Monthly, Seasonal and Annual Temperature Reconstructions for Central Europe Derived from Documentary Evidence and Instrumental Records Since AD 1500. *Climatic Change*, Vol.101, No.1-2, pp. 69-107, ISSN 1573-1480

Eddy, A. (1976). The Maunder Minimum. *Science*, Vol. 192, No.4245, pp.1189-1202, ISSN 1095-9203

Eichkorn, S.; Wilhelm, S.; Aufmhoff, H.; Wohlfrom, K. H. & Arnold, F. (2002). Cosmic Ray-Induced Aerosol-Formation: First Observational Evidence from Aircraft-Based Ion Mass Spectrometer Measurements in the Upper Troposphere. *Geophysical Research Letters*, Vol. 29, No.14, 1698, DOI: 10.1029/2002GL015044, ISSN 0094-8276

Esper, J.; Cook, E.R. & Schweingruber, F.H. (2002). Low-Frequency Signals in Long Tree-ring Chronologies for Reconstructing Past Temperature Variability. *Science*, Vol. 295, No.5563 , pp. 2250-2253, ISSN 1095-9203

Farquhar, G.D.; O'Leary, M.H. & Berry, J.A. (1982). On the relationship between carbon isotope discrimination and the intercellular carbon dioxide concentration in leaves. *Australian Journal of Plant Physiology*, Vol.9, pp.121-137, ISSN 0310-7841

Ferretti, M.; Innes, J.L.; Jalkanen, R.; Saurer, M.; Schäffer, O.; Spiecker, H. & Wilpert, K. (2002). Air pollution and environmental chemistry what role for tree-ring studies? *Dendrochronologia*, Vol. 20, No.1-2, pp. 159-174, ISSN 1125-7865

Forster, P. M. & Gregory, J. M. (2006). The Climate Sensitivity and Its Components Diagnosed from Earth Radiation Budget Data. *Journal of Climate*, Vol.19, No.1, pp. 39–52. doi:10.1175/JCLI3611.1., ISSN 0894-8755

Frame, D.J.; Booth, B.B.B.; Kettleborough, J.A.; Stainforth, D.A.; Gregory, J.M.; Collins, M. & Allen, M.R. (2005). Constraining Climate Forecasts: the Role of Prior Assumptions. *Geophysical Research Letters*, Vol.32, L09702, ISSN 0094-8276

Friis-Christensen, E. & Lassen, K. (1991). Length of Solar Cycle: an Indicator of Solar Activity Closely Associated With Climate. *Science*, Vol.254, No5032, pp. 698-700, ISSN 1095-9203

Fritts, H. (1976). *Tree rings and climate*, Academic Press, London

Fröhlich, C. & Lean, J. (1998).The Sun's Total Irradiance: Cycles and Trends in the Past Two Decades and Associated Climate Change Uncertainties. *Geophysical Research Letters*, Vol. 25, No.23 , pp. 4377– 4380, ISSN 0094-8276

Gagen, M.; Zorita, E.; McCarroll, D.; Young, G.H.F.; Grudd, H.; Jalkanen, R.; Loader, N.J.; Robertson, I. & Kirchheler, A. (2011). Cloud Response to Summer Temperatures in Fennoscandia over the Last Thousand Years. *Geophysical Research Letters*, Vol.38, L05701. doi: 10.1029/2010GL046216, ISSN 0094-8276

Gray, L. J.; Beer, J.; Geller, M.; Haigh, J.D.; Lockwood, M.; Matthes, K.; Cubasch, U.; Fleitmann, D.; Harrison, G.; Hood, L.; Luterbacher, J.; Meehl, G. A.; Shindell, D.; van Geel B. & White, W. (2010). Solar influences on climate. *Reviews in Geophysics*, Vol.48, RG4001, doi:10.1029/2009RG000282, ISSN 8755-1209

Harrison, R.G. & Ambaum, M.H.P. (2009). Observed Atmospheric Electricity Effect on Clouds. *Environmental Research Letters*, Vol.4, 014003, ISSN 1748-9326

Helama, S.; Timonen, M.; Lindholm, M.; MeriläInen, J. & Eronen, M. (2005). Extracting Long-Period Climate Fluctuations from Tree-Ring Chronologies Over Timescales of Centuries to Millennia. *International Journal of Climatology*, Vol. 25, pp.1767-1779, ISSN 0899-8418

Herron, M. (1982). Impurity Sources of F, Cl, NO_3 and SO_4 in Greenland and Antarctic Precipitation. *Journal of Geophysical Research*, Vol.87, No.C4, pp. 3052–3060, ISSN 0094-8276

Hilasvuori, E.; Berninger , F.; Sonninen, E.; Tuomenvirta, H. & Jungner, H. (2009). Stability of Climate Signal in Carbon and Oxygen Isotope Records and Ring Width From Scots Pine (*Pinus sylvestris* L.) in Finland. *Journal of Quaternary Science*, Vol.24, No.5, pp.469-480, ISSN 0267-8179

Hoyt, D.V. & Schatten, K.H. (1993). A Discussion on Plausible Solar Irradiance Variations, 1700-1992 . *Journal of Geophysical Research*, Vol. 98, No.A11, pp. 18895-18906, ISSN 0094-8276

Hoyt, D. & Schatten, K. H. (1998). Group Sunspot Numbers: a New Solar Activity Reconstruction. *Solar Physics*, Vol.179, pp. 189-219, ISSN 0038-0938

Idso, S.B. (1998). CO_2-Induced Global Warming: a Skeptic's View of Potential Climate change. *Climate Research*, Vol. 10, pp. 69 – 82, ISSN 0936-577x

IPCC (2007). WGI Fourth Assessment Report: Climate Change 2007: *The Physical Science Basis: Summary for Policymakers*, Paris

Jalkanen, R.; Tuovinen, M.; (2001). Annual Needle Production and Height Growth: Better Climate Predictors Than Radial Growth at Treeline? *Dendrochronologia*, Vol.19, No, pp.39–44, ISSN 1125-7865

Jokipii, J.R. (1991). Variation of the cosmic-ray flux in time, In: *The Sun in time*. C.P. Sonnett, M.S. Giampapa & M.S. Matthews, (Eds.), 205-220, Tuscon, University of Arizona press

Jones, P.D.; Briffa, K.R.; Barnett, T.P. & Tett, S.F.B. (1998). High-Resolution Palaeoclimatic Records for the Last Millennium: Interpretation, Integration and Comparison with General Circulation Model Control-Run Temperatures. *The Holocene*, Vol. 8.4, pp. 455-471, ISSN 0959-6836

Jones, P.D., Parker, D.E., Osborn, T.J. & Briffa, K.R. (2001). Global and hemispheric temperature anomalies - land and marine records. In: *Trends: A compendium of data on global change*, Carbon dioxide information analysis center, Oak Ridge National Laboratory, US Department of Energy, Oak Ridge, Tennessee, USA

Johnsen, S.J.; Dansgaard, W.; Clausen, H.B. & Langway, C.C. (1972). Oxygen Isotope Profiles Through the Antarctic and Greenland Ice Sheets. *Nature*, Vol.235, No.5339, pp.429 - 434, ISSN 0028-0836

Kazil, J. & Lovejoi, E.R.(2004). Tropospheric ionization and aerosol production: a model study. *Journal of Geophysical Research*. Vol. 109. D19206, doi:10.1029/2004JD004852, ISSN 0148-0227

Kocharov, G.E.; Ostryakov, V.M.; Peristykh, A.N. and Vasil'ev, V.A. (1995). Radiocarbon Content Variations and Maunder Minimum of Solar Activity. *Solar Physics*, Vol. 159, pp. 381-395.

Kocharov, G.E.; Akhmetkereev, S.Kh. & Peristykh, A.N. (1990). On the solar flare effect in the atmospheric radiocarbon. In: *Possibilities of the methods of measurement of ultra-minor amount of isotopes*, G.E. Kocharov, (Ed.), 45-70, PhTI, Leningrad (in Russian)

Kocharov, G.E.; Ogurtsov, M.G. & Dreschoff, G.A.M. (1999). On the Quasi-Five-Year Variation of Nitrate Abundance in Polar Ice and Solar Flare Activity in the Past. *Solar Physics*, Vol.188, pp.187-190, ISSN 0038-0938

Kocharov, G.E.; Koudriavtsev, I.V.; Ogurtsov, M. G.; Sonninen, E. & Jungner, H. (2000). The Nitrate Content of Greenland Ice and Solar Activity. *Astronomy Reports*, Vol. 44, No. 12, pp. 825-829, ISSN 1063-7329

Konstantinov, B.P. & Kocharov, G E. (1965). *Astrophysical Phenomena and Radiocarbon*. Doklady Akademii Nauk SSSR, Vol.165, pp. 63–64, ISSN 0869-5652 (in Russian)

Kostantinov, A.N.; Levchenko, V.A.; Kocharov, G.E.; Mikheeva, I.B.; Cecchini, S.; Galli, M.; Nanni, T.; Povinec, P.; Ruggiero, L. & Salomoni, A. (1992). Theoretical and Experimental Aspects of Solar Flares Manifestation in Radiocarbon Abundance in Tree Rings. *Radiocarbon*, Vol. 34, No. 2, pp. 247-253, ISSN 0033-8222

Kurt, V.; Belov, A.; Mavromichalaki, H.; & Gerontidou, M. (2004). Statistical analysis of solar proton events. *Annales Geophysicae*, Vol.22, No.6, pp.2255-2271, doi:10.5194/angeo-22-2255-2004, ISBN 0992-7689

Kondratyev, K.Ya. & Nikolsky, G.A. (1983). The Solar Constant and Climate. *Solar Physics*, Vol. 89, pp. 215-222, ISSN 0038-0938

Kondratyev, K.Ya. (1999). *Climatic effects of aerosol and clouds*. Springer/Praxis U.K., Chichester

Laken, B. A.; Kniveton, D. R. & Frogley, M. R. 2010. Cosmic Rays Linked to Rapid Mid-Latitude Cloud Changes. *Atmospheric 1 and Physics Discussions*, Vol. 10, pp. 18235–18253, ISSN 1680-7367

Lean, J.; Beer J. & Bradley, R. (1995). Reconstruction of Solar Irradiance since 1610: Implications for Climate Change. *Geophysical Research Letters*, Vol. 22, No.23, pp. 3195-3198, ISSN 0094-8276

Lee, S.-H.; Reeves, J. M.; Wilson, J. C.; Hunton, D. E.; Viggiano, A. A.; Miller, T. M.; Ballenthin, J. O. & Lait, L. R. (2003). Particle Formation by Ion Nucleation in the Upper Troposphere and Lower Stratosphere. *Science*, Vol.301, No.5641, pp.1886-1889, ISSN1095-9203

Legrand, M.R.; Stordal, F.; Isaksen, I.S.A. & Rognerud, B. (1989). A Model Study of the Stratospheric Budget of Odd Nitrogen, Including Effects of Solar Cycle Variations. *Tellus*, Vol. 41B, pp. 413-426, ISSN

Legrand, M.R. & Kirchner, S. (1990). Origins and Variation of Nitrate in South Polar Precipitations. *Journal of Geophysical Research*, Vol. 95, pp. 3493-3507, ISSN 0148-0227

Lindzen, R.S. & Giannitsis, C. (1998). On the climatic implications of volcanic cooling. *Journal of Geophysical Research*, Vol. 103, No. D6, pp. 5929-5941, ISSN 0148-0227

Lindzen, R.S.; Chou, M.-D. & Hou, A.Y. (2001). Does the Earth Have an Adaptive Infrared Iris? *Bulletin of the American Meteorological Society*, Vol. 82, pp. 417-432, ISSN 1520-0477

Lindzen, R. S. & Choi, Y.-S. (2009). On the Determination of Climate Feedbacks from ERBE Data. *Geophysical Research Letters*. Vol. 36, L16705, doi:10.1029/2009GL039628, ISSN 0094-8276

Loehle, C.A. (2007). 2000-year global temperature reconstruction based on non-treering proxies. *Energy and Environment*, Vol. 18, No. 7-8, pp.1049-1058, ISSN 0958-305X

Logan, J.A. (1983). Nitrogen Oxides in the Troposphere: Global and Regional Budgets. *Journal of Geophysical Research*, Vol.88, No.C15, pp. 10785-10805, ISSN 0148-0227

Lu, H.; Jarvis, M.J.; Graf, H.-F.; Young, P.C. & Horne, R.B. (2007) Atmospheric Temperature Responses to Solar Irradiance and Geomagnetic Activity. *Journal of Geophysical Research*, Vol. 112, No. D11, D11109, doi:10.1029/2006JD007864, ISSN 0148-0227

Mann, M.E.; Bradley, R.S. & Hughes, M.K. (1999). Northern Hemisphere Temperatures During the Past Millennium: Inferences, Uncertainties, and Limitations. *Geophysical Research Letters*, Vol. 26, No.6, pp. 759-762, ISSN 0094-8276

Mann, M. & Hughes, M. (2002). Tree Ring Chronologies and Climate Variability. *Science*, Vol.296, No.5569, pp. 848-852, ISSN 1095-9203

Mann, M.E.; Rutherford, S.; Bradley, R.S.; Hughes, M.K. & Keimig, F.T. (2003). Optimal Surface Temperature Reconstructions Using Terrestrial Borehole Data. *Journal of Geophysical Research*, Vol. 108, No.D7, 4203, doi:10.1029/2002JD002532, ISSN 0148-0227

Marichev, V.N.; Bogdanov, V.V.; Zhivet'ev, I.V. & Shvetsov, B.M. (2004). Influence of the Geomagnetic Disturbances on the Formation of Aerosol Layers in the Stratosphere. Geomagnetism and Aeronomy, Vol.44, No.6, pp. 779-786, ISSN 0016-7932

Marsh, N. & Svensmark, H. (2000). Low Cloud Properties Influenced by Cosmic Rays. *Physical Review Letters*, Vol. 85, No. 23, pp. 5004-5007, ISSN 0031-9007

Mayewski, P.A.; Meeker, L.D.; Morrison, M.C.; Twickler, M.S.; Whitlow, S.; Ferland, D. A.; Legrand, M.R. & Steffenson, J. (1993). Greenland Ice Core «Signal» Characteristics:

an Expanded View of Climate Change. *Journal of Geophysical Research,* Vol. 98, No.D7, pp. 12839-12847, ISSN 0148-0227

Mayewski, P.A.; Lyons, W.B.; Twickler, M.S.; Buck, C.F. & Whitlow, S. (1990). An Ice Record of Atmospheric Response to Antropohenic Sulphate and Nitrate. *Nature,* Vol. 346, No. 6284, pp. 554-556., ISSN 0028-0836

Mc Cracken, K. G.; Dreschhoff, G. A. M.; Smart, D. F. & Shea, M.A. (2001A). Solar Cosmic Ray Events for the Period 1561-1994: 2. The Gleissberg periodicity. *Journal of Geophysical Research,* Vol.106, No.A10, pp. 21599-21610, ISSN 0148-0227

Mc Cracken, K.G.; Dreschhoff, G.A.M.; Zeller, E.J.; Smart, D.F. & Shea, M.A. (2001B). Solar Cosmic Ray Events for the Period 1561-1994. 1. Identification in Polar Ice, 1561-1950. *Journal of Geophysical Research,* Vol. 106, No. A10, pp. 21585-21598, ISSN 0148-0227

Mc Carroll, D.; Jalkanen, R.; Hicks, S.; Tuovinen, M.; Gagen, M.; Paweliek, F.; Eckstein, D.; Schmitt, U.; Autio, J. & Heikkinen, O. (2003). Multiproxy dendroclimatology: a pilot study in northern Finland. *The Holocene,* Vol.13, No.6, pp. 829-838, ISSN 0959-6836

Moberg, A.; Sonechkin, D.M.; Holmgren, K.; Datsenko, M.M. & Karlen, W. (2005). High Variable Northern Hemisphere Temperatures Reconstructed from Low- and High-Resolution Proxy Data. *Nature,* Vol. 433, No.7026, pp. 613-617, ISSN 0028-0836

Mordvinov, A.V.; Makarenko, N.G.; Ogurtsov, M.G. & Jungner, H. (2004). Reconstruction of magnetic activity of the Sun and changes in its irradiance on a millennium timescale using neurocomputing. *Solar Physics,* Vol. 224, pp. 247-253, ISSN 0038-0938

Nagovitsyn, Yu.A. (1997). A Nonlinear Mathematical Model for the Solar Cyclicity and Prospects for Reconstructing the Solar Activity in the Past. *Astronomy Letters,* Vol. 23, No. 6, pp. 742 - 748, ISSN 1063-7737

Nagovitsin, Yu.A.; Ivanov, V.G.; Miletsky, E.V. & Volobuev, D. (2003). Solar Activity Reconstruction from Proxy Data. In.: *Proc. of International conference-workshop "Cosmogenic climate forcing factors during the last millennium".* 41-50, Vytautas Magnus University, ISBN 9955-530-89-8, Kaunas

Nagovitsyn, Yu.A.; Ivanov, V.G.; Miletsky Kaunas, E.V. & Volobuev, D.M. (2004). ESAI Data Base and Some Properties of Solar Activity in the Past. *Solar Physics,* Vol. 224, No. 1-2, pp. 103-112, ISSN 0038-0938

Nagovitsyn, Yu.A. (2005). To the Description of Long-Term Variations in the Solar Magnetic Flux: the Sunspot Area Index. *Astronomy Letters,* Vol.31, No.8, pp. 557-562, ISSN 1063-7737

Ney, E.P. (1959). Cosmic radiation and weather. *Nature,* Vol.183, pp. 451-452, ISSN 0028-0836

Nicoll, K.A., & Harrison, R.G. (2009). Vertical Current Flow through Extensive Layer Clouds. *Journal of Atmospheric and Solar-Terrestrial Physics,* Vol. 71, No. 12, pp. 1219-1221, ISSN 1364-6826

Obridko, V.N. (2008). Magnetic Fields and Active Complexes. In: *Plasma HelioGeoPhysics monograph,* L.M. Zelenyi and I.S. Veselovsky (Eds.), 41-60, Fizmatlit, ISBN 978-5-9221-1040-2, Moscow, (in Russian)

Ogurtsov, M.G.; Kocharov, G.E.; Lindholm, M.; Meriläinen, J.; Eronen, M. & Nagovitsyn, Yu.A (2002). Evidence of Solar Variation in Tree-Ring-Based Climate Reconstructions. *Solar Physics,* Vol. 205, No.2, P. 403-417, ISSN 0038-0938

Ogurtsov, M.G.; Jungner, H.; Kocharov, G.E.; Lindholm M. & Eronen, M. (2004). Nitrate Concentration in Greenland Ice: an Indicator of Changes in Fluxes of Solar and Galactic High-Energy Particles. *Solar Physics*, Vol.222, pp.177-190, ISSN 0038-0938

Ogurtsov, M.G. (2005). On the Possibility of Forecasting the Sun's Activity Using Radiocarbon Solar Proxy. *Solar Physics* , Vol.231, No.1-2, pp.167-176, ISSN 0038-0938

Ogurtsov, M.G. & Lindholm, M. (2006). Uncertainties in Assessing Global Warming During the 20th Century: Disagreement Between Key Data Sources. *Energy and Environment*, Vol.17, No.5, pp. 685-706, ISSN 0958-305X

Ogurtsov, M. G. (2007). Secular Variation in Aerosol Transparency of the Atmosphere as the Possible Link Between Long-Term Variations in Solar Activity and Climate. *Geomagnetism and Aeronomy*, Vol. 47, No. 1, pp. 118-128, ISSN 0016-7932

Ogurtsov, M.G. (2010). Long-Term Solar Cycles According to Data on the Cosmogenic Beryllium Concentration in Ice of Central Greenland. *Geomagnetism and Aeronomy*, Vol. 50, No. 4, pp. 475–481, ISSN 0016-7932

Ogurtsov, M.G.; Jungner, H.; Helama, S.; Lindholm, M. & Oinonen, M. (2011). Paleoclimatological Evidence for Unprecedented Recent Temperature Rise at the Extratropical Part of the Northern Hemisphere. *Geografiska Annaler*, Vol. 93, No.1, pp.17-27, ISSN 1468-0459

Ogurtsov, M.G. (2011). Relationship Between the Aerosol Concentration in the Stratosphere and Ionization According to the Data on Conductivity and Nitrate Content of Greenland Ice. *Geomagnetism and Aeronomy*, Vol.51, No.2, pp. 267-274, ISSN 0016-7932

Palle, E.; Buttler, C.J. & O'Brien, K. (2004). The Possible Connection Between Ionization in the Atmosphere by Cosmic Rays and Low Level Clouds. *Journal of Atmospheric and Solar-Terrestrial Physics*, Vol. 66, No, pp. 1779–1790, ISSN 1364-6826

Palmer, A.S.; van Ommen, T.D.; Curran, A.J. & Morgan, V. (2001). Ice-Core Evidence for a Small-Source of Atmospheric Nitrate. *Journal of Geophysical Research,*. Vol. 28, No.10, pp. 1953-1956, ISSN 0148-0227

Pensa, M.; Sepp, M. & Jalkanen, R. (2006). Connections Between Climatic Variables and the Growth and Needle Dynamics of Scots Pine (Pinus sylvestris L.) in Estonia and Lapland. *International Journal of Biometeorology*, Vol.50, pp.205–214, ISSN 0020-7128

Pudovkin, M.I. (1996). Effect of the Solar Activity on the Lower Atmosphere and Weather. *Soros Educational Journal*, No.10, pp. 106-114 (in Russian)

Pudovkin, M.I. & Veretenenko, S.V. (1992). Effect of Geomagnetic Disturbances on the Intensity of Direct Solar Radiation Flux. *Geomagnetism and Aeronomy*, Vol.32, No.1 , pp. 114 –115, ISSN 0016-7932

Roden, J.S.; Lin, G. & Ehleringer, J.R. (2000). A Mechanistic Model for Interpretation of Hydrogen and Oxygen Ratios in Tree-Ring of Cellulose. *Geochimica et Cosmochimica Acta*, Vol.64, pp.21-35, ISSN 0016-7037

Roldugin, V.K.& Vashenyuk, E.V. (1994). Atmospheric Transparency Variations Caused by Cosmic Rays. *Geomagnetism and Aeronomy*, Vol.34, No.2, p. 251-253, ISSN 0016-7932

Roldugin, V.C. & Tinsley, B.A.(2004). Atmospheric Transparency Changes Associated with Solar Wind-Induced Atmospheric Electricity Variations. *Journal of Atmospheric and Solar-Terrestrial Physics*, Vol.66, No.13-14, pp. 1143–1149, ISSN 1364-6826

Schove, D.J (1983). *Sunspot Cycles*. Hutchinson Ross Publications, ISBN 0-87033-424-X, Stroudsburg Pennsylvania

Schwartz, S. E. (2008). Reply to comments by G. Foster et al., R. Knutti et al., and N. Scafetta on Heat capacity, time constant, and sensitivity of Earth's climate system. *Journal of Geophysical Research*, Vol.113, D15105, doi:10.1029/2008JD009872, ISSN 0148-0227

Seo, J.-W.; Eckstein, D.; Jalkanen, R. & Schmitt, U. (2011). Climatic Control of Intra- and Inter-Annual Wood-Formation Dynamics of Scots Pine in Northern Finland. *Environmental and Experimental Botany*, doi: 10.1016/ienvexpbot.2Q 11.01.003, (in press).

Shindell, D.; Rind, D.; Balachandran, N.; Lean & P. Lonergan. (1999). Solar Cycle Variability, Ozone and Climate. *Science*, Vol.284, No, pp. 305-308, ISSN 1095-9203

Shumilov, O.I.; Kasatkina, E.A.; Henriksen, K. & Vashenyuk, E.V.(1996). Enhancement of Stratospheric Aerosol after Solar Proton Event. *Annales Geophysicae*, Vol. 4, No.11, pp. 1119-1123, ISSN 0992-7689

Shvedov, P.N. (1892). Tree as archive of droughts. *Meteorological bulletin*, No.5 (in Russian).

Solanki, S.; Usoskin, I.G.; Kromer, B.; Schüssler, M. & Beer, J. (2004). Unusual Activity of the Sun During Recent Decades Compared to the Previous 11,000 years. *Nature*, Vol. 431, No.7012, pp. 1084 - 1087, ISSN 0028-0836

Starkov, G.V. & Roldugin, V.K. (1995). Relationship Between Atmospheric Transparency Variations and Geomagnetic Activity. *Geomagnetism and Aeronomy*, Vol. 34, No. 4, p. 559-562, ISSN 0016-7932

Stozhkov, Y. I. (2003). The Role of Cosmic Rays in the Atmospheric Processes. *Journal of Physics G: Nuclear Particle Physics*, Vol.29, pp.913–923 PII: S0954-3899(03)54503-0, ISSN 0954-3899

Stuiver, M.; Reimer, P. J.; Bard, E.; Beck, J.W.; Burr, G.S.; Hughen, K.A.; Kromer, B.; McCormac, G.; van der Plicht, M. & Spurk, M. (1998). INTCAL98 Radiocarbon Age Calibration, 24 000-0 cal BP. *Radiocarbon*, Vol. 40, No.3, pp. 1041-1083, ISSN 0033-8222

Solanki, S. (2003). Sunspots: An Overview. *The Astronomy and Astrophysics Review*, Vol.11, pp.153–286, ISSN 0935-4956

Svensmark, H.; Pedersen, J.O.;.Marsh, N. M.; Enghoff, ,M.B. & Uggerhøj U.I. (2007). Experimental Evidence for the Role of Ions in Particle Nucleation Under Atmospheric Conditions. Proceedings of the Royal Society A . Vol. 463, pp. 385-396, ISSN 1364-5021

Svensmark, H.; Bondo, T. & Svensmark, J. (2009). Cosmic Ray Decreases Affect Atmospheric Aerosols and Clouds. *Geophysical Research Letters*, Vol.36, L15101, doi:10.1029/2009GL038429, ISSN 0094-8276

Tinsley, B. A., Deen, G.W. (1991). Apparent Tropospheric Response to MeV-GeV Particle Flux Variations: a Connection via Electrofreezing of Supercooled Water in High-Level Clouds. *Journal of Geophysical Research*, Vol. 96, No.12, pp. 22283-22296, ISSN 0148-0227

Tinsley, B.A. (2000), Influence of solar wind on the global electric circuit, and inferred effects on cloud microphysics, temperature, and dynamics in the troposphere. *Space Science Review*, Vol.94, pp.231-258, ISSN 0038-6308

Tinsley, B.A. (2008). The Global Atmospheric Electric Circuit and its Effects on Cloud Microphysics. *Reports on Progress in Physics*, Vol. 71, 066801, ISSN 0034-4885

Todd, M. & Kniveton, D.R. (2004). Short-Term Variability in Satellite-Derived Cloud Cover and Galactic Cosmic Rays: an Update. *Journal of Atmospheric and Solar-Terrestrial Physics*, Vol. 66, No.13-14, pp. 1205-1211, ISSN 1364-6826

Tuovinen, M.; Mc Carroll, D.; Grudd, H.; Jalkanen, R. & Los, S. (2009). Spatial and Temporal Stability of the Climatic Signal in Northern Fennoscandian Pine Tree-Ring Width and Maximum Density. *Boreas*, Vol.38, No.1, pp. 1-12, ISSN 1502-3885

Urey, H.C. (1948). Oxygen Isotopes in Nature and in the Laboratory. *Science*, Vol.108, No.2813. pp. 602-603., ISSN 1095-0203

Usoskin, I.G.; Mursula, K.; Solanki, S.; Schüssler, M. & Alanko, K. (2004). Reconstruction of Solar Activity for the Last Millennium Using 10Be Data. *Astronomy and Astrophysics*, Vol.413, pp.745–751, ISSN 0004-6361

Usoskin, I.G.; Solanki, S.K. & Taricco, C. (2006). Long-Term Solar Activity Reconstructions: Direct Test by Cosmogenic ^{44}Ti in Meteorites. *Astronomy and Astrophysics*, Vol. 457, L25-28, ISSN 0004-6361

Usoskin, I.G.; Solanki, S.; Kovaltsov, G.A.; Beer, J. & Kromer, B. (2006). Solar Proton Events in Cosmogenic Isotope Data. *Geophysical Research Letters*, Vol.33, No. 8, L08107, ISSN 0094-8276

Vasiliev, S.S. & Dergachev, V.A. (2002). The ~2400-year cycle in atmospheric radiocarbon concentration: bispectrum of ^{14}C data over the last 8000 years. *Annales Geophysicae*. Vol. 20, pp. 115-120, ISBN 0992-7689

Veretenenko, S. V. & Pudovkin M. I. (1993). Effects of Cosmic Ray Variations in the Circulation of the Lower Atmosphere. *Geomagnetism and Aeronomy*, Vol. 33, No.6, pp. 35-40, ISSN 0016-7932 (in Russian).

Veretenenko, S. V.; Dergachev, V. A. & Dmitriev, P. B. (2007). Effect of Solar Activity and Cosmic-Ray Variations on the Position of the Arctic Front in the North Atlantic. *Bulletin of the Russian Academy of Sciences: Physics*, Vol.71, No.7, pp. 1010-1013, ISSN 1062-8738

Vitinsky, Y.I.; Kopecky, M. & Kuklin, G.V. (1986). *Statistics of Sunspot Activity*, Nauka, Moscow, USSR (in Russian)

Wild, G. (1882). *On the Air temperature in Russian Empire*. No. 2, St.Petersburg, (in Russian)

Willson, R.C. & Mordvinov, A.V. (2003). Secular solar irradiance trend during solar cycles 21-23. *Geophysical Research Letters*, Vol. 30, doi:10.1029/2002GL016038, ISSN 0094-8276

Young, G. H. F.; McCarroll, D; Loader, N. J. & Kirchhefer, A. (2010). A 500-year Record of Summer Near-Ground Solar Radiation from Tree-Ring Stable Carbon Isotopes. *The Holocene*, Vol.20, No.3, pp. 315-324, ISSN

Yu, F. & Turco, R. P. (2000). Ultrafine Aerosol Formation via Ion-Mediated Nucleation. *Geophysical Research Letters*, Vol.27, No, pp. 883-886, ISSN 0094-8276

Yu, F. & Turco, R. P. (2001). From Molecular Clusters to Nanoparticles: The Role of Ambient Ionization in Tropospheric Aerosol Formation. *Journal of Geophysical Research*, Vol.106, No, pp. 4797-4814, ISSN 0148-0227

Yu, F. (2006). From Molecular Clusters to Nanoparticles: Second-Generation Ion-Mediated Nucleation Model. *Atmospheric Chemistry and Physics Discussions*, Vol. 6, pp. 3049–3092, ISSN 1680-7367

Zeller, E.J. & Parker, B.C. (1981). Nitrate Ions in Antarctic Firn as a Marker of Solar Activity. *Geophysical Research Letters*, Vol. 8, No.8, pp. 895-898, ISSN 0094-8276

Part 3

Climate and ENSO

4

Assimilating Ocean Observation Data for ENSO Monitoring and Forecasting

Yosuke Fujii[1], Masafumi Kamachi[1], Toshiyuki Nakaegawa[1], Tamaki Yasuda[1], Goro Yamanaka[1], Takahiro Toyoda[1], Kentaro Ando[2] and Satoshi Matsumoto[3]

[1]Japan Meteorological Agency/ Meteorological Research Institute

[2]Japan Agency for Marine-Earth Science and Technology

[3]Japan Meteorological Agency

Japan

1. Introduction

El Niño–Southern Oscillation (ENSO) is one of the most influential fluctuations of the coupled atmosphere-ocean climate system in the Seasonal-to-Interannual (SI) time scale for global and local communities. Ocean data assimilation systems are generally adopted for monitoring ENSO because the variation of the heat content in the ocean interior of the equatorial Pacific is considered a good precursor of El Niños, and essential for understanding the ENSO process. They are also utilized for the initialization of Coupled Atmosphere-Ocean General Circulation Models (CGCMs), along with atmosphere data assimilation systems in the SI forecasting systems of various operational centers.

In this paper, we discuss the capacity of current operational ocean data assimilation systems adopted for ENSO monitoring and SI forecasting based on studies using the system of the Japan Meteorological Agency (JMA). It then demonstrates the benefits of assimilating ocean observation data through those systems in SI forecasts using the JMA seasonal and ENSO forecasting system. It also introduces the recent effort of JMA/Meteorological Research Institute (MRI) to resolve "coupled shock", which is one of the most crucial issues concerning the initialization with uncoupled ocean and atmosphere data assimilation systems in SI forecasting.

This chapter is organized as follows. Section 2 summarizes the process of developing ocean data assimilation systems for up-to-date ENSO monitoring and SI forecasting. In particular, it describes efforts to improve the Temperature-Salinity (T-S) balance in the assimilation results over the past decade. We then introduce the ocean data assimilation system used in the JMA seasonal and ENSO forecasting system as a state-of-the-art system in Section 3. Section 4 demonstrates the importance of assimilating ocean observation data through the ocean data assimilation system for ENSO and seasonal forecasting. Section 5 introduces the recent effort to resolve the coupled shock at JMA/MRI. This chapter is summarized in Section 6.

2. First and second generations of ocean data assimilation systems

There have been two innovative efforts in the development of ocean data assimilation systems for ENSO monitoring and SI forecasting, and both of them occurred following material changes of the ocean observing system.

The first effort followed the deployment of the Tropical Atmospheric Ocean (TAO) array (Hayes et al., 1991; McPhaden et al., 1998) under the Tropical Ocean and Global Atmosphere (TOGA) program (e.g., Ji et al., 1995). This change initially realized the operational monitoring of the ocean interior state in the equatorial Pacific using an ocean data assimilation system. However, since major oceanic data observed by Autonomous Temperature Line Acquisition System (ATLAS) buoys (mooring buoys that form the TAO array) were temperature profiles, ocean data assimilation systems developed in this early stage adopted the Optimal Interpolation (OI) or Three-Dimensional Variational Method (3DVAR) assimilating only temperature profiles. Salinity profiles were abandoned in those systems, although their number was not large. It should also be noted that these data assimilation systems directly modified only the model temperature fields; the model salinity field was modified only through adjustment by model physics. Development of schemes assimilating Sea Surface Height (SSH) data started after the launch of the TOPEX/Poseidon satellite, but it still focused on modifying the model temperature field for the first several years (e.g., Ji et al., 2000).

Some studies, however, pointed out that appropriate treatment of salinity is essential for controlling model fields realistically, for the following two reasons. One is the effect of salinity variations on SSH and pressure fields through contribution to density variations. This issue was addressed by Cooper (1988); Ji et al. (2000); Maes (1998). In particular, Ji et al. (2000) demonstrated that the use of observed SSH in addition to in situ temperature data can increase analysis errors in the systems analyzing temperature alone without considering the salinity anomaly, because the spurious temperature anomaly is estimated in order to compensate the contribution of salinity to SSH.

The other reason is the possibility of destroying density stratification, which was first addressed by Woodgate (1997). Temperature and salinity decrease with depth in many areas of the tropical and subtropical oceans. The upward shift of water mass induces cold and low salinity anomalies in those areas. When assimilating temperature alone, the cold anomaly is reproduced but the low salinity anomaly is missed, resulting in the simulated greater density exceeding the actual density. This artificial density tends to destabilize the stratification and to induce spurious vertical mixing. This effect severely diminishes the salinity maximums in the subsurface layers in the tropical oceans (e.g., Troccoli et al., 2002).

Many studies have also stressed the importance of salinity variability in the equatorial Pacific. The surface salinity front existing in the western equatorial Pacific is considered a good indicator of the eastern edge of the warm water pool, and its position is directly connected to the advective-reflective oscillator theory (e.g., Picaut et al., 1997). Roemmich et al. (1994) suggested that the salinity front affects the current fields in the near-surface layer.

Formation of the barrier layer (the isothermal layer in which salinity is stratified; Lukas & Lindström, 1991) is another important feature associated with salinity there. This barrier layer can affect surface currents by thinning the mixed layer and concentrating the effect of wind stress (e.g., Vialard et al., 2002). It is also considered to induce a temperature rise in the mixed layer because it prevents warm water at the surface from mixing with cold water in the thermocline. This tendency is confirmed in the equatorial Pacific (e.g., Ando & McPhaden, 1997; Fujii et al., 2012; Maes et al., 2006). Some studies (e.g., Maes & Belamari, 2011) have further used CGCMs to confirm the impact of the barrier layer at the onset of El Niños.

In addition, the number of Argo floats rapidly increased after 2000, in order to meet the requirements of the Global Ocean Data Assimilation Experiment (GODAE) project (Clark et al., 2009). This second material change in the observing system urged to develop schemes for assimilating observed salinity profiles effectively in ocean data assimilation systems. In 2000, the name of the TAO array was also changed to TAO/Triangle Trans-Ocean Buoy Network (TRITON) array to designate the upgrading of ATLAS buoys west of 160°E to TRITON buoys that contain more enriched equipment (Kuroda, 2002). However, many in situ observation platforms still exist, including the TAO/TRITON array, which provide little salinity data.

In order to use those observation data effectively along with the data of simultaneous observations of temperature and salinity by Argo floats and other sophisticated platforms, it is necessary to develop a method to guarantee consistency between temperature and salinity fields, even when only a small amount of salinity observation data exist. Thus, for the past decade, various schemes estimating salinity fields mainly from temperature and SSH through the T-S balance relationship have been developed based on OI or 3DVAR (summarized in Fujii et al., 2010).

For example, salinity fields are estimated using regression coefficients of their anomaly with the SSH anomaly, as well as temperature fields by Ezer & Mellor (1994); Kamachi et al. (2001). Some studies (e.g., Carton et al., 2000; Huang et al., 2008; Yan et al., 2004) uses climatological T-S relations typically calculated from the World Ocean Atlas (e.g., Locarnini et al., 2010). Vertical shifts of water masses in background (model) T-S profiles are applied in Haines et al. (2006); Ricci et al. (2005). Coupled T-S Empirical Orthogonal Function (EOF) modal decomposition is also employed to reconstruct vertical temperature and salinity profiles in Dobricic et al. (2005); Fujii & Kamachi (2003); Maes et al. (2000).

Thus, most current operational ocean data assimilation systems adopt OI or 3DVAR and have the capacity to assimilate observed salinity profiles imposing a multivariate (mainly T-S) balance relationship. These systems are called "second-generation" systems in Balmaseda et al. (2009; 2011), while systems that can assimilate temperature data alone are called "first-generation" systems. Balmaseda et al. (2011) demonstrated an example of improved ENSO forecasting from the first generation to the second generation. In the following section, we also confirm the advantage of second-generation systems.

It should be noted that data assimilation systems adopting more sophisticated schemes (e.g., ensemble Kalman filter or the adjoint method) have also been developed or started to use in operation (e.g., Keppenne et al., 2005; Weaver et al., 2003). These systems have the potential to improve the accuracy of monitoring and forecasting skills further.

3. Capacity of the second-generation system at JMA

This section introduces MOVE/MRI.COM-G (Usui et al., 2006, hereafter, MOVE-G) as a typical second-generation ocean data assimilation system, and discusses the capacity improvement of second-generation systems based on assimilation runs of MOVE-G. This system is also employed in the studies of Sections 4 and 5.

MOVE-G is composed of an Ocean General Circulation Model (OGCM) based on the MRI Community Ocean Model (MRI.COM; Tsujino et al., 2010), and the Multivariate Ocean Variational Estimation System (MOVE), an analysis scheme based on Fujii & Kamachi (2003). The OGCM has a nearly global domain extending from 75°S to 75°N. The grid spacing is 1° in the zonal direction, and changes from 0.3° (within 5.7°S - 5.7°N) to 1° (poleward of 16°S and 16°N) in the meridional direction. It has 50 levels in the vertical direction. The first seven levels are 1, 3, 6, 10, 16, 22, and 30 m in depth, and the interval between 30 and 200 m depth is

10 m. A generalized enstrophy-preserving scheme and a scheme that involves the concept of diagonally upward/downward mass momentum fluxes along the sloping bottom (Ishizaki & Motoi, 1999) are applied for momentum advection. Isopycnal diffusion (Redi, 1982), isopycnal thickness diffusion (Gent & McWilliams, 1990), and the vertical mixing scheme of Noh & Kim (1999) are also adopted.

The analysis scheme MOVE adopts the 3DVAR method using vertical coupled T-S EOF modal decomposition for the background error covariance matrix. In this scheme, correction of temperature and salinity profiles to their first guess values, $\Delta\mathbf{x}_p$, is represented by a linear combination of the T-S EOF modes as follows:

$$\Delta\mathbf{x}_p = \sum_l w_l \lambda_l \mathbf{S}\mathbf{u}_l \,, \tag{1}$$

where w_l is the amplitude of the lth EOF mode, λ_l is the singular value for the lth mode, $\mathbf{S}$ is the diagonal matrix composed of the standard deviations of temperature and salinity from the first guess, and $\mathbf{u}_l$ is the vector representing the lth mode. It should be noted that the vector $\Delta\mathbf{x}_p$ contains the corrections of temperature and salinity at all model levels as its elements.

In MOVE-G, the model domain is partitioned into 40 horizontal subdomains, and the T-S EOF modes are calculated for each subdomain from historical profile data in the World Ocean Database 2001 (WOD01; Conkright et al., 2002) and the Global Temperature-Salinity Profile Program (GTSPP) database (Hamilton, 1994). The subdomains overlap in boundary areas. Correction of temperature and salinity profiles in the overlapping areas is calculated as a weighted sum,

$$\mathbf{x}_p = \mathbf{x}_p^f + \sum_m a_m \Delta\mathbf{x}_{p,m} \,, \tag{2}$$

where $\mathbf{x}_p$ is the analysis of the profiles, $\mathbf{x}_p^f$ is the first guess of the profiles, m is the counter of the subdomains, $\Delta\mathbf{x}_{p,m}$ is the correction based on the EOF modes of the mth subdomain, and a_m is the weight of the mth subdomain that satisfies $\sum_m a_m^2 = 1$ for avoiding the loss or gain of total variances by the area partition (Fukumori, 2002).

The mode amplitudes, w_l in (1), are determined to minimize the cost function. The cost function, $J(\mathbf{w})$, is defined as

$$\begin{aligned} J(\mathbf{w}) = & \frac{1}{2}\sum_m\sum_l \mathbf{w}_{m,l}^T \mathbf{B}_m^{-1} \mathbf{w}_{m,l} + \frac{1}{2}(\mathbf{H}\mathbf{x}-\mathbf{y})^T \mathbf{R}^{-1}((\mathbf{H}\mathbf{x}-\mathbf{y})) \\ & + \frac{1}{2}[\mathcal{H}_h(\mathbf{x})-\mathbf{y}_h]^T \mathbf{R}_h^{-1}[\mathcal{H}_h(\mathbf{x})-\mathbf{y}_h] + J_{add} \,, \end{aligned} \tag{3}$$

where $\mathbf{w}$ is the vector containing the amplitudes at all horizontal grid points for all subdomains, and $\mathbf{w}_{m,l}$ is the partial vector of $\mathbf{w}$ containing the amplitudes of the lth mode in the mth subdomain. The matrix $\mathbf{B}_m$ represents the horizontal correlation of background (first guess) errors for the mth subdomain modeled by the Gaussian function. The vector $\mathbf{y}$ is composed of temperature and salinity observations, and $\mathbf{y}_h$ is satellite SSH data. Vector $\mathbf{x} = \mathbf{x}^f + \mathbf{G}\mathbf{w}$ is the state vector of temperature and salinity analysis fields composed of $\mathbf{x}_p$, where $\mathbf{x}^f$ is the first guess and $\mathbf{G}$ denotes the transformation represented by (1) and (2). Matrix $\mathbf{H}$ represents spatial interpolation for acquiring the values equivalent to the temperature and salinity observation, and $\mathcal{H}_h$ is the nonlinear operator that includes calculating the Sea surface Dynamic Height (SDH) from gridded temperature and salinity data and interpolation. Matrix $\mathbf{R}$ ($\mathbf{R}_h$) is the error covariance matrix for the temperature and salinity profiles (satellite SSH

data). The term J_{add} is the additional nonlinear constraint for avoiding density inversion (Fujii et al., 2005).
The gradient of the cost function is written as

$$\mathbf{g} = \sum_m \sum_l \mathbf{B}_m^{-1} \mathbf{w}_{m,l} + \mathbf{G}^T \left[\mathbf{H}^T \mathbf{R}^{-1} (\mathbf{H}\mathbf{x} - \mathbf{y}) + \mathbf{H}_h{}^* \mathbf{R}_h{}^{-1} \{ \mathcal{H}_h(\mathbf{x}) - \mathbf{y}_h \} \right] + \nabla J_{add}, \quad (4)$$

where $\mathbf{H}_h^*$ is the adjoint code of the nonlinear operator $\mathcal{H}_h$. We also apply the variational quality control procedure introduced in Fujii et al. (2005). The cost function is non-quadratic for $\mathbf{w}$, and the calculation of $\mathbf{g}$ includes inversion of the non-diagonal matrix $\mathbf{B}_m$. In MOVE-G, we adopt the preconditioned quasi-Newton method introduced in Fujii (2005) to minimize this non-quadratic function without directly implementing the inversion.
The 3DVAR analysis explained above is performed once every assimilation cycle using all available observations in the term of the cycle, and the result is reflected in the model fields by Incremental Analysis Updates (IAU: Bloom et al., 1996). The first guess for the analysis is given as a weighted mean of the climatology and the model-prediction for the middle time of the cycle from the assimilation result at the end of the previous cycle. The difference between the analysis and the model-prediction (i.e., analysis increment) is applied to correct the temperature and salinity fields in the model. Current fields are adjusted to the corrected temperature and salinity fields through the model dynamics, and thus establish the geostrophic balance in most areas.
An online model-bias estimation using the one-step bias-correction algorithm (Balmaseda et al., 2007) can be applied with IAU in MOVE-G. The bias estimates are subtracted from the model-prediction fields before calculating the first guess, and are updated by taking a weighted mean of its original and analysis increment in every assimilation cycle.
In addition, the SSH change due to the variation of the total fresh water mass in the global ocean, which is not taken into account in the OGCM, is estimated in the assimilation runs in Section 4. In the estimation, this globally constant value is regarded as a control variable of 3DVAR. This value is added to SDH calculated from the temporal analysis fields before subtracting the observed SSH. The term of the background constraint for the SSH change is also added to the cost function (3). The estimation in the analysis in the previous assimilation cycle is adopted as the first guess. We set a small value for the prescribed error variance of the first guess in order to vary the value slowly.
In situ temperature and salinity profiles, satellite SSH anomaly data, and observation-based gridded Sea Surface Temperature (SST) data are assimilated into the model in MOVE-G. The temperature and salinity profiles employed in the assimilation runs in this study are collected from WOD01, GTSPP, and the data of the TAO/TRITON array. Profiles of Argo floats are included in GTSPP. The SSH data is the along-track data from TOPEX/Poseidon, Jason-1, ERS-1/2, and ENVISAT, extracted from Ssalto/Duacs delayed-time multimission altimeter products (CLS, 2004). Centennial in-situ Observation-Based Estimates of the variability of SST and marine meteorological variables (COBE-SST; Ishii et al., 2005), or the gridded SST data compiled in JMA, are also assimilated in the assimilation runs in Sections 4 and 5.
In the previous section, we indicated that state-of-the-art (second-generation) operational ocean data assimilation systems adopt schemes in which consistency between temperature and salinity is assumed. MOVE-G also improves the accuracy of the salinity fields by establishing an adequate relationship between temperature and salinity through the coupled T-S EOF modes. In particular, variation of salinity coupled with that of temperature can be

estimated through the coupled T-S EOF modes, even if little salinity data is available (Fujii & Kamachi, 2003).

Here, we compare the temperature and salinity fields in three assimilation runs in order to demonstrate the difference between the first- and second-generation ocean data assimilation systems. One is the assimilation run named MOVE-G VAL in Fujii et al. (2012). The atmospheric reanalysis dataset produced by the National Center for Environmental Prediction and the National Center for Atmospheric Research (NCEP-R1; Kalnay et al., 1996) is employed as the external forcing in MOVE-G VAL. Online model-bias estimation is applied in this assimilation run. The length of the assimilation cycle is set to one month. The gridded SST data are not assimilated.

In addition, several profiles of temperature and salinity are excluded from the assimilated data in order to use them as independent reference data. The excluded profiles are those observed by TRITON buoys positioned at 5°N-156°E, 0°-156°E, and 5°S-156°E; the profiles of Argo floats whose last digit of the World Meteorology Organization (WMO) ID is "4"; and the profiles whose position and date are similar to those of one of the above excluded profiles (within 0.1° in longitude and latitude, and on the same date).

In another assimilation run, MOVE-G 1GE, only temperature observations are assimilated, and salinity increments estimated from the temperature data through T-S EOFs are not applied to correct the model salinity field. Thus, the model salinity field is just adjusted to the corrected temperature field through the model physics in MOVE-G 1GE. This run is equivalent to those of the first-generation systems, while MOVE-G VAL can be considered an assimilation run of the second-generation system. Observation data other than temperature are not assimilated in the other run, MOVE-G 2GE-T, either; however, the salinity increments calculated through T-S EOFs are applied to correct the model salinity fields there.

The settings of the assimilation system for MOVE-G 1GE and MOVE-G 2GE-T, other than those described above, are the same as those for MOVE-G VAL. The observation data withheld in MOVE-G VAL are also withheld in MOVE-G 1GE and MOVE-G 2GE-T. We analyze the results of these assimilation runs in the period of 1993–2008.

Validation of salinity using the data observed by TRIRON buoys deployed along 156°E (Fig. 1) clearly indicates that using T-S EOFs improves the salinity field even if salinity observations are not assimilated. The subsurface salinity maximum is apparently diminished at 0° and 5°S in MOVE-G 1GE. The variations of near-surface fresh water at these positions are not estimated satisfactorily in this run. The subsurface salinity maximum is smoothed out vertically at 5°N. These errors stem from the spurious vertical mixing induced by the breakdown of the T-S relation due to modifying the temperature field alone (Troccoli et al., 2002).

The subsurface low salinity biases at 0° and 5°S are removed, and the variations of the surface fresh water are fairly estimated in MOVE-G 2GE-T due to the T-S EOFs. The subsurface salinity at 5°N is also improved, although the contrast of the subsurface salinity between 0° and 5°N is slightly underestimated.

Assimilating salinity and SSH improves the salinity field further. The subsurface salinity contrast between 0° and 5°N is improved, and appearance of water whose salinity exceeds 35.4 at 0° is properly estimated in MOVE-G VAL. It should be noted that the observation data of these TRIRON buoys are not assimilated as well in MOVE-G VAL, as in the other two assimilation runs.

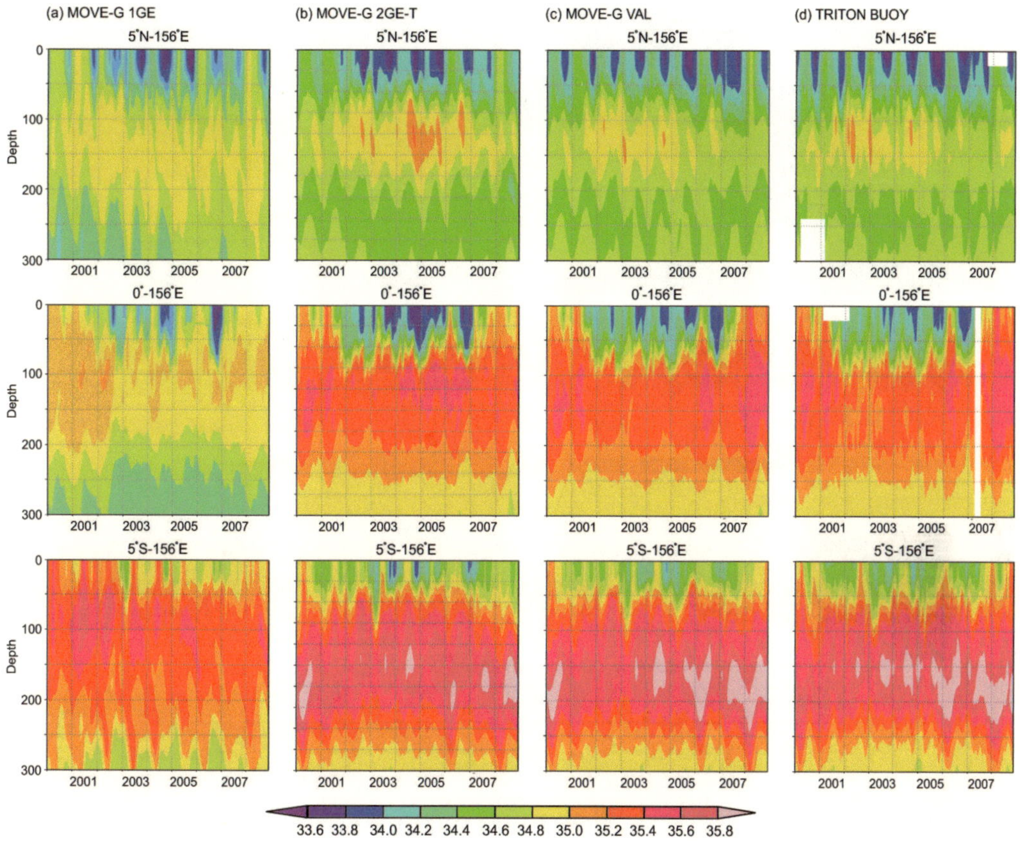

Fig. 1. Comparison of salinity in (a) MOVE-G 1GE, (b) MOVE-G 2GE-T, and (c) MOVE-G VAL, with (d) salinity profiles observed by TRITON buoys at 5°N-156°E (top), 0°-156°E (middle), and 5°S-156°E (bottom).

Figure 2(a) indicates that MOVE-G 1GE has a colder temperature field than MOVE-G VAL in the entire equatorial Pacific. This colder temperature is also induced by spurious vertical mixing due to modification of only the model temperature field.

We further calculate statistics for the accuracy of temperature fields in the equatorial Pacific for MOVE-G 1GE and MOVE-G VAL. The reference for the statistics is the data of the profiling floats in 2°S-2°N, 130°E-80°W that are withheld in the assimilation runs. The result (Table 1) reveals that spurious mixing actually causes a cold bias.

Table 1 also indicates that imposing a T-S balance relationship in the second-generation system resolves this problem. The difference between the mean temperature field of MOVE-G 2GE-T and that of MOVE-G VAL (Fig. 2(b)) also implies that imposing the T-S relationship effectively suppresses spurious mixing and reduces the cold bias.

Smaller Root Mean Square Differences (RMSDs) and larger Anomaly Correlation Coefficients (ACCs) suggest that not only the mean state but also the variability of the temperature field in the equatorial Pacific is improved in MOVE-G VAL. The accuracy of temperature and salinity fields in the equatorial Pacific is thus improved in second-generation systems, as a result of imposing the T-S balance relationship.

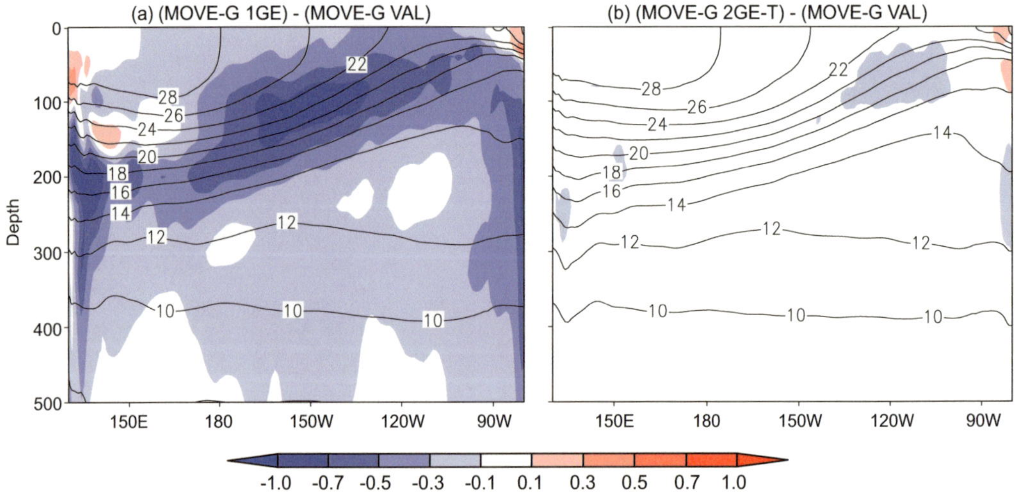

Fig. 2. Mean temperatures (contour) of (a) MOVE-G 1GE and (b) MOVE-G 2GE-T in 1993–2008 and their differences from the mean of MOVE-G VAL (shading) in the Pacific equatorial section. Units in °C.

	Bias (°C)		RMSD (°C)		ACC	
Depth	1GE	VAL	1GE	VAL	1GE	VAL
10m	-0.134	0.029	0.572	0.487	0.795	0.828
50m	-0.120	0.079	0.620	0.553	0.860	0.880
100m	-0.471	0.030	2.386	1.121	0.561	0.645
150m	-0.584	0.120	1.893	1.536	0.522	0.651
200m	-0.350	0.129	1.391	1.202	0.447	0.553
300m	-0.259	-0.116	0.557	0.410	0.247	0.548

Table 1. Statistics associated with the accuracy of temperatures at the fixed depths for MOVE-G 1GE (1GE) and MOVE-G VAL (VAL) in the equatorial Pacific. BIAS and RMSD: Averaged difference and RMSD between 1GE/VAL and reference values. ACC: Correlation coefficient between anomalies of 1GE/VAL and reference values. Anomalies are calculated as the deviation from the World Ocean Atlas 2009 (Locarnini et al., 2010). Values equivalent to the reference for the assimilation runs are calculated by spatial and temporal interpolations from the monthly mean data.

4. Impact of ocean observation data on ENSO forecasting

In this section, we demonstrate the benefits of assimilating ocean observation data through the ocean data assimilation system in ENSO and seasonal forecasting. The impact of oceanic temperature and salinity observed by the TAO/TRITON array and profiling floats (most deployed as Argo floats) on the SI forecasting is examined in order to show the benefits.

Temperature profiles of the TAO array were clearly the most influential data on ENSO forecasting before 2000. However, the recent increase of Argo floats is likely to reduce the influence of the TAO/TRITON array, since both observe the subsurface temperature field in the tropical Pacific.

Possibly, oceanic temperature and salinity information from the TAO/TRITON array and profiling floats is mostly redundant. If so, part of the cost for the observing platforms is abandoned. Therefore, showing the complementary effects of the array and the floats is vital for sustaining these observation platforms. In particular, the impact of the observing platforms on SI forecasting is important information for administrators of the observing system, because SI forecasting is one of the most influential products of ocean observation data.

The complementary impacts of TAO/TRITON array and profiling floats have already been demonstrated by an Observing System Experiment (OSE) using the seasonal forecasting system of the European Centre for Medium-Range Weather Forecasts (ECMWF) (Balmaseda & Anderson, 2009). The results of OSEs, however, depend greatly on the forecasting system. Therefore, we also perform OSEs continuously in order to examine the impact of assimilating oceanic temperature and salinity data of the TAO/TRITON array and profiling floats through MOVE-G in the JMA seasonal and ENSO forecasting system. It should be noted that this study does not evaluate the atmospheric data and oceanic current data of the TAO/TRITON array. Our previous results are briefly introduced in Balmaseda et al. (2009). In this section, we introduce our recent OSE results.

The OSE configuration is as follows. First, we perform three ocean data assimilation runs (ALL, XTT, and XAF) from 2000 to 2008 using MOVE-G. All available data in the regular observation dataset are assimilated in ALL. Data of the TAO/TRITON array are withheld in XTT, and data of the profiling floats are withheld in XAF.

The three assimilation runs use the same setting except for the assimilated data. The period of the assimilation cycle is 10 days, the same as in JMA's operational system. The Japanese 25-year Reanalysis (JRA-25; Onogi et al., 2007) is used for estimating the atmospheric forcing. The original period of JRA-25 is 1979 to 2004, but it is extended to the present time by adding the product of the JMA Climate Data Assimilation System (JCDAS). The online model-bias estimation is not applied here. However, we applied the estimation of the SSH change due to the variation of the total fresh water mass in the global ocean. COBE-SST is assimilated in all the assimilation runs.

Thirteen-month 11-member ensemble forecasts are then performed with JMA/MRI-CGCM, the CGCM adopted in the forecasting system (Takaya et al., 2010; Yasuda et al., 2007), from 31 January, 26 April, 30 July, and 28 October during the 5 years of 2004–2008. We thus have 20 forecasts. Analysis fields of ALL, XTT, and XAF are used as the initial conditions of the ocean component in these forecasts. The atmospheric fields of JRA-25 are used for the initial condition of the atmospheric component. The flux correction procedure used in the JMA operation is applied in all forecast calculations.

In order to generate ensemble members, we put small perturbations on the observation data of the gridded SST in the 10-day assimilation cycle just before the start of a forecast. Although we applied perturbations in the same shape as the regression map of SST to the NINO4 index (see Table 2), we found in a preliminary study that forecast results are not sensitive to the shape of the perturbation if the scale is as small as we applied in this study.

We calculate the ensemble mean of horizontal monthly mean fields for several atmospheric and oceanic parameters from all the forecast results and convert them to data in a 2.5°-interval grid, although only the results for SST, Sea Level Pressure (SLP), Velocity Potential on the 200 hPa surface (VP200), and Outgoing Longwave Radiation (OLR) are presented here. Forecast biases are estimated by averaging the deviations of forecast values from their references separately for each lead time; each forecast month; and each of ALL, XTT, and XAF. The forecast values are then calibrated by subtracting the corresponding biases. Thus, each set of forecasts for ALL, XTT, and XAF is separately calibrated. It should be noted that a forecast value is employed to calibrate itself. Although this procedure is not fair for evaluating a forecast skill, we adopt it because of the shortage of forecasts. We assume this procedure does not affect the conclusion of this study. We use COBE-SST for the reference of SST, the National Oceanic and Atmospheric Administration (NOAA) OLR (Liebmann & Smith, 1996) for the reference of OLR, and JRA-25 for the reference of the other atmospheric parameters (SLP and VP200).

NINO1+2	10°S-0°,	90-80°W	NINO3	5°S-5°N,	150-90°W
NINO3.4	5°S-5°N,	170-120°W	NINO4	5°S-5°N,	160°E-150°W
TRITON	5°S-5°N,	125-160°E	Philippine	5-25°N,	125-150°E
IODE	10-0°S,	90-110°E	IODW	10°S-10°N,	50-70°E

Table 2. Definition of the areas in which the SST anomaly is averaged for calculating SST indices.

First, we analyze the impact of the TAO/TRITON array and profiling floats on the forecasts of the area-averaged SST indices. The indices are calculated by averaging the anomaly of monthly mean SST from the monthly climatology of the reference data in the areas defined in Table 2. The impact of the TAO/TRITON array (profiling floats) is evaluated through the difference in ACCs with the reference between the forecasts from ALL and XTT (XAF) (i.e., the increase of ACC if the data of the array (floats) are assimilated in addition to the other data in the regular observation dataset).

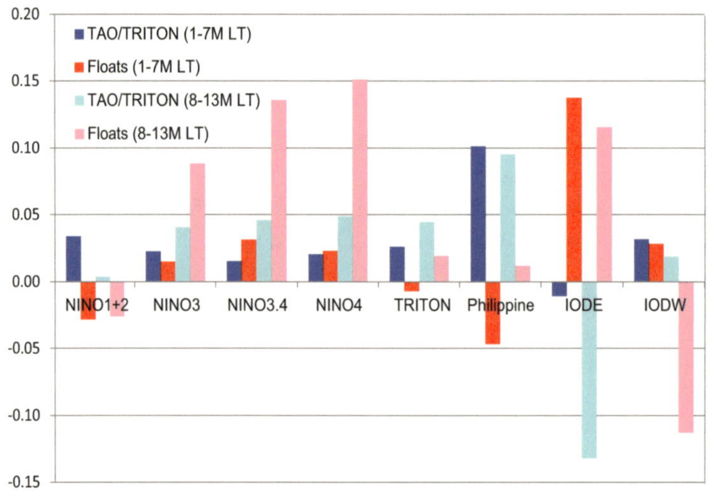

Fig. 3. Difference in ACCs with the reference between the ensemble mean forecasts of area-averaged SST indices from ALL and XTT or XAF. Blue: ALL−XTT for 1–7M LT forecasts. Red: ALL−XAF for 1–7M LT forecasts. Light blue: ALL−XTT for 8–13M LT forecasts. Pink: ALL−XAF for 8–13M LT forecasts. ACCs for 1–7M (8–13M) LT forecasts are calculated from all forecasted results of 1, 2, · · · , 7 (8, 9, · · · , 13) month lead times. Blue and light blue (red and pink) bars denote the impacts of the TAO/TRITON array (profiling floats).

Figure 3 indicates that the TAO/TRITON array improved the SST forecasts in the areas other than IODE. Data of the array increases ACCs by 0.02 for the 1–7 Month Lead Time (1–7M LT) forecasts of the SST indices of the equatorial Pacific (NINO1+2, NINO3, NINO3.4, NINO4, and TRITON). The impact of the array increases and reaches more than 0.04 for 8–13 Month Lead Time (8–13M LT) forecasts in the equatorial Pacific, except NINO1+2. The TAO/TRITON array also has a large impact on SST in the Philippine Sea. It improves the ACC for the Philippine SST index by about 0.1. Although it decreases the ACC for 8–13M LT forecasts of SST in the eastern Indian Ocean (IODE) by more than 0.1, it has a positive impact for the western Indian Ocean (IODW).

Figure 3 also indicates that 8–13M LT forecasts of SST are greatly improved by assimilating the float data in the eastern and central equatorial Pacific, except NINO1+2. The increase of ACC is more than 0.1 for NINO3.4 and NINO4, and about 0.09 for NINO3, although its impact is as large as that of the TAO/TRITON array for 1–7M LT forecasts in these areas. The longer lead time ENSO forecasts are likely to be effectively improved by the better subsurface temperature fields in the whole tropical Pacific, due to the assimilation of the float data.

Assimilating float data has a negative impact on NINO1+2. The impact is also negative in the western tropical Pacific (the TRITON and Philippine area) for 1–7M LT forecasts although it becomes positive for 8–13M LT forecasts there. Assimilating float data also increases ACCs of SST in the eastern Indian Ocean (IODE) more than 0.1 and improves ACC of the western Indian Ocean (IODW) about 0.03 for 1–7M LT forecasts. Although the large impact on IODE remains for 8–13M LT forecasts, the skill of forecasting the SST index for IODW is severely degraded for 8–13M LT forecasts.

Thus, the oceanic data of both the TAO/TRITON array and profiling floats generally have positive impacts on the SST forecasts in the equatorial Pacific. Here, the positive impacts of these oceanic data can be regarded as complementary because they mean that the forecast skills are improved by assimilating TAO/TRITON (float) data in addition to the float (array) and other regular data. They also suggest that adding temperature and salinity profiles to the current regular observation data (including TAO/TRITON and float data) may further improve the forecast skills.

These positive impacts also affect the forecast skills of the atmospheric state. Figure 4 indicates the impact of assimilating data of the TAO/TRITON array or the profiling floats on SLP, VP200, and OLR for 1-7M LT forecasts. Forecasts of SLP fields are remarkably improved by assimilating TAO/TRITON array in the central and eastern tropical Pacific, with a direct link to the forecast improvement of the SST anomaly that indicates EL Niños and La Niñas. It also improves SLP in the Indian Ocean and south of Japan probably due to the remote effects of ENSO.

VP200 is also improved by TAO/TRITON data in a wide area, particularly over North America and the area extending from northern China to the Philippine Sea. Figures 4(d, e) indicate that assimilating float data also has a positive impact, very similar to that of assimilating TAO/TRITON data, on the forecasts of SLP and VP200.

The increase of ACC for VP200 is caused by better representation of the vertical air mass transports associated with precipitation in the tropics. Figures 4(c, f) indicate that assimilating the data of the TAO/TRITON array or profiling floats improves the variation of OLR, which is considered a proxy of precipitation, in a wide area over the Pacific including the western subtropical North Pacific south of Japan, around the maritime continent and Australia, and in the tropical and subtropical South Indian Ocean. Improvement of the divergence fields in the upper troposphere (VP200) leads to better upper-tropospheric wind fields and adequate representation of the global-scale atmospheric circulation. Thus, increasing the amount of

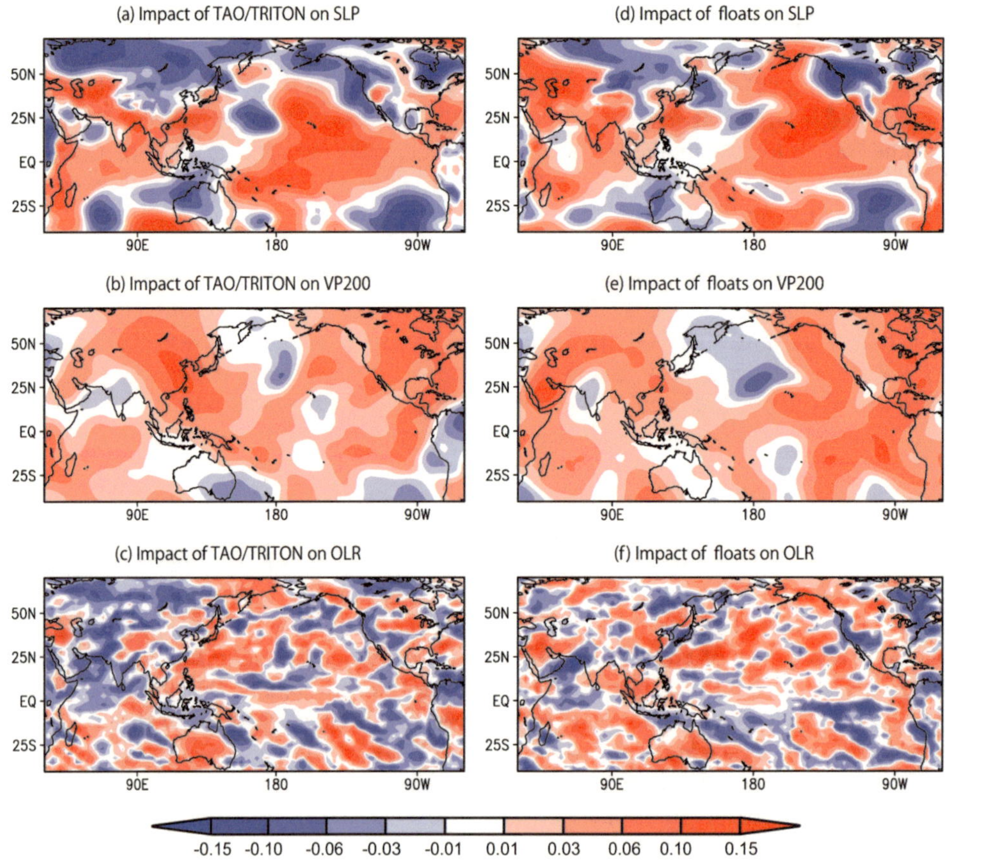

Fig. 4. Maps of the difference in ACCs with the reference between 1-7M LT ensemble mean forecasts of SLP, VP200, and OLR from ALL and XTT or XAF. (a) ALL−XTT for SLP. (b) ALL−XTT for VP200. (c) ALL−XTT for OLR. (d) ALL−XAF for SLP. (e) ALL−XAF for VP200. (f) ALL−XAF for OLR. ALL−XTT (ALL−XAF) represents the impact of the TAO/TRITON array (profiling floats).

ocean data for assimilation has a possibility to improve the forecast skills of the atmospheric state globally.

It should, however, be noted again that the impact of observation data in an OSE is highly dependent on the forecasting system (i.e., the forecasting model and the data assimilation scheme). In addition, it depends on the target phenomena of the OSE. Therefore, we should examine the impacts of ocean observation data on variety of targets using various forecasting systems in order to evaluate the values of the data appropriately and to optimize the ocean observing system.

For that purpose, delayed-mode OSEs of ocean observation data for weekly to decadal time scales, including SI forecasts, are encouraged by the observing system evaluation task team of the GODAE Ocean View project (Oke et al., 2009; 2011, see also https://www.godae-oceanview.org/science/task-teams/observing-system-evaluation-tt-oseval-tt/). The task team plans to compile the results of OSEs and release them as the observation impact statement in order to share the values of ocean observations with the public and the administrators of observation platforms.

5. Constraining a coupled model using ocean data assimilation: an effort to resolve the coupled shock

In the previous section, we demonstrated the benefits of assimilating ocean observation data through ocean data assimilation systems in ENSO and seasonal forecasting. However, it should be noted that ocean analysis/reanalysis fields calculated by ocean assimilation systems generally have some inconsistency with atmospheric analysis/reanalysis fields because they are not calculated simultaneously in a unified data assimilation system. This inconsistency induces so-called the "coupled shock" when the ocean and atmospheric fields are used as the initial condition of a CGCM in an ENSO and seasonal forecasting system, and possibly reduces the improvement of the forecast skill due to the assimilation of the ocean observation data.

An essential method of resolving the coupled shock is developing a coupled data assimilation system, that is, a system in which atmosphere and ocean observation data are employed to constrain a CGCM. Information in the ocean observation data are extracted more effectively in a coupled data assimilation system, because it can improve not only oceanic but also atmospheric analysis/reanalysis fields.

Contemplating the benefits described above, coupled data assimilation systems were developed by Zhang et al. (2007), based on an ensemble Kalman filter, and by Sugiura et al. (2008) based on an adjoint method. However, developing a coupled data assimilation system requires tremendous human effort and a heavy computer burden. The difference in the major time-scale between the atmosphere and ocean is another challenge.

In order to deal with the coupled shock, we developed a "quasi-coupled data assimilation system" that uses only ocean observation data to constrain the ocean component of a CGCM (atmospheric observation data are not assimilated). This system is named MOVE-C. MOVE-C adopts the coupled model for JMA's SI forecasting, JMA/MRI-CGCM, and we apply the same procedure of the ocean data assimilation scheme as in MOVE-G since the oceanic part of the CGCM is the same as the OGCM adopted in it. We assume that slow components (i.e., climate variabiliy) can be subtracted from the full variability of the coupled atmosphere and ocean system by assimilating only ocean observation data in MOVE-C. It is a proto type of a truly coupled data assimilation system that we intend to develop in the future.

In order to examine the feasibility of the quasi-coupled data assimilation system, we conduct a five-member ensemble assimilation run in the period of 1979–2008 using MOVE-C. Here, we apply the online model-bias estimation, the length of the assimilation cycle is set to one month, and COBE-SST is assimilated. It should be noted that a single-member assimilation run was used in the analysis in Fujii et al. (2009). We assume that using a five-member ensemble increases the reliability of the analysis, compared to the previous study.

We also conduct a five-member ensemble of the Atmospheric Model Intercomparison Project (AMIP) runs, or simulation runs of the atmosphere model used in MOVE-C with the observed daily SST data (COBE-SST) as the oceanic boundary condition, in the same period. In addition, we use a free simulation run of JMA/MRI-CGCM, the CGCM used in MOVE-C. The simulation started from an assimilation result of MOVE-G and JRA-25 on 1 January 2000, and the integration is performed for 101 years. We use the simulation result of the last 60 years here. We also use COBE-SST, NOAA OLR, and JRA-25 as the reference for SST, OLR

and the other atmospheric parameters. All datasets are converted to monthly mean with a grid spacing of 2.5° before calculating statistics and indices in this section.
In the previous study, we demonstrated that the precipitation field was improved in the tropics in a single-member assimilation run of MOVE-C over an AMIP run. First, we reconfirm the improvement using the five-member ensemble run. In order to compare the accuracy of the MOVE-C with that of the AMIP run, we calculate ACCs between OLR (a proxy of precipitation) in those runs and reference data.

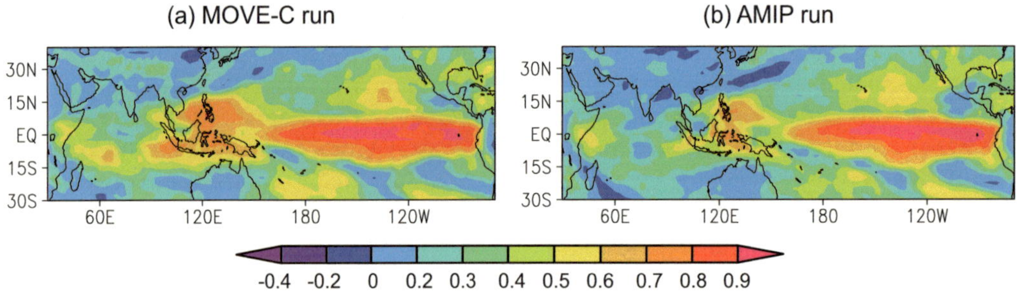

Fig. 5. ACC maps of ensemble mean OLR with the reference data in (a) MOVE-C run and (b) AMIP run. ACCs are calculated from anomalies in all months in 1980–2007. The three-month running mean of the deviation from the monthly climatology of each run or reference is adopted as the anomaly.

Figure 5 indicates that ACC for OLR is improved in the western equatorial Pacific and Philippine Sea, around the maritime continent, and in a wide area of the Indian Ocean. ACC is also increased around the Himalayas. This result is consistent with that of the previous study (Fujii et al., 2009).
In the previous study, we suggested that the precipitation field is deteriorated with the absence of negative feedback between SST and precipitation in the AMIP run. In the real world, warm (cold) SST tends to increase (decrease) precipitation, while enhanced (suppressed) convection tends to induce an SST drop (rise) because of the cloud cover and the condition of ocean mixing. This negative feedback does not work appropriately in the AMIP run because SST is prescribed.
Our analysis suggests that this defect of the AMIP run is likely to suppress the atmospheric response to ENSO. Figure 6 presents the distribution of simultaneous correlation coefficients of the anomaly of VP200 with the NINO3 index. It should be noted that the precipitation field is directly coupled with the vertical air mass transportation and therefore with the divergence in the upper troposphere (a negative value of VP200).
In an El Niño period (when the NINO3 index is positive), increased precipitation in the central equatorial Pacific induces a divergent anomaly in the upper troposphere (negative anomaly of VP200) there, while decreased precipitation in the western equatorial Pacific and maritime continent induces a convergent anomaly (positive anomaly of VP200) there. Thus, the correlation coefficients are positive around the maritime continent and negative over the central equatorial Pacific.
Comparison of the correlation coefficients of the AMIP run (Fig. 6(b)) with those of the reference data (Fig. 6(a)) indicates that both negative and positive correlations are underestimated. This weak response of the atmosphere to ENSO probably occurs for the following reason. Precipitation (a drought) tends to decrease (increase) SST in the real world. However, precipitation is likely to be underestimated (overestimated) if the decreased

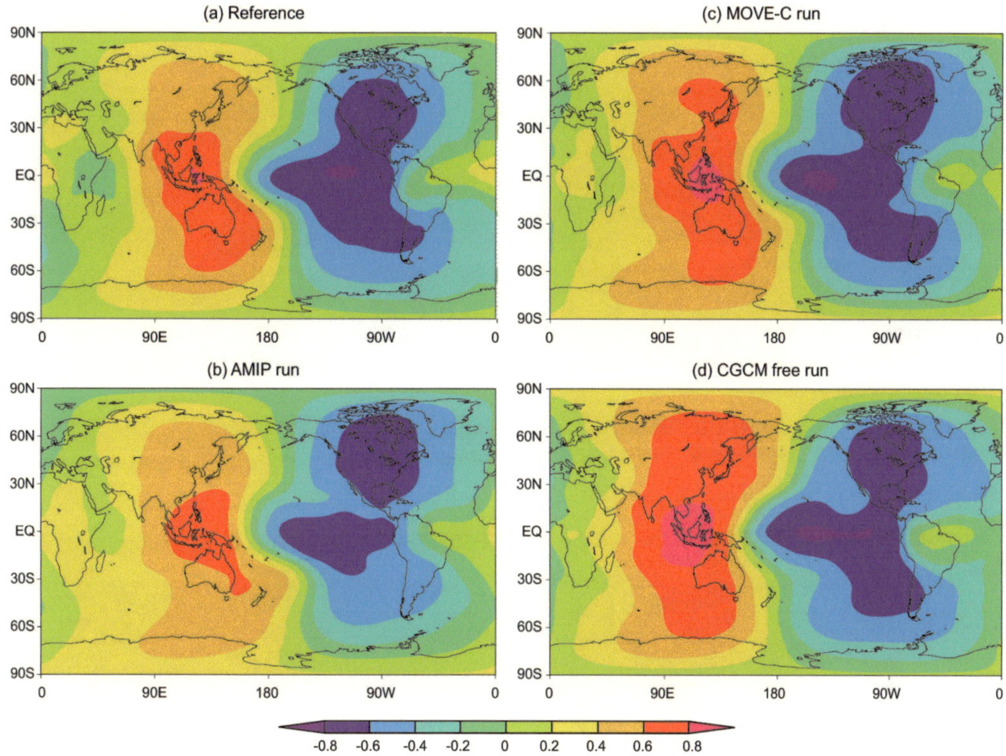

Fig. 6. Maps of the simultaneous correlation coefficient of the anomaly of VP200 from its monthly climatology with the NINO3 index in (a) reference data, (b) AMIP run, (c) MOVE-C run, and (d) CGCM Free run. The coefficient is calculated for the period of 1980–2007. The anomaly is the deviation from the monthly climatology, and the three-month running mean is applied for both VP200 and the index. The correlation coefficients averaged for ensemble members are adopted for the MOVE-C and AMIP runs.

(increased) SST is given as the oceanic boundary condition in the AMIP run. Thus, the response of the atmosphere to SST through precipitation is suppressed in the AMIP run.

This problem is resolved by taking into account the air-sea interaction through a CGCM. If the decreased (increased) SST suppresses (enhances) precipitation, the model increases (decreases) SST by the negative feedback, and augments (reduces) precipitation. The precipitation is thus kept at the appropriate level. Figure 6(c) confirms that the negative correlation of VP200 with the NINO3 index over the central Pacific is adequately reproduced in the MOVE-C run, and the positive correlation over the maritime continent is improved in the run over the AMIP run, although it is slightly overestimated reflecting the inherent property of the CGCM that is implied by the map for the CGCM free run (Fig. 6(d)). The overestimated response to ENSO in the free run may be caused by the absence of part of the natural variability due to lacks of some model physics.

The improved response of the atmosphere to ENSO is likely to improve the variability of VP200 (Fig. 7). In the MOVE-C run, the area where ACC exceeds 0.8 spreads wider than in the AMIP run over the maritime continent. ACCs also apparently increase over the western Indian Ocean, East Asia, western North Pacific, and east of the Hawaii Islands.

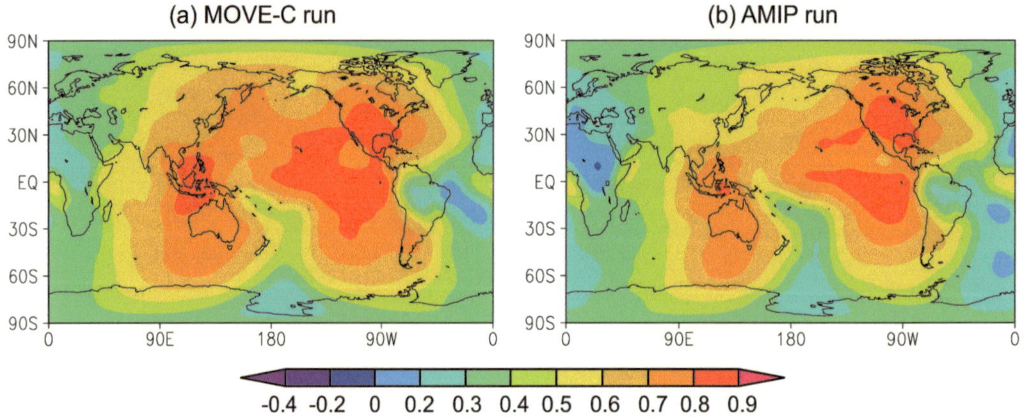

Fig. 7. Same as Fig. 5 but for VP200.

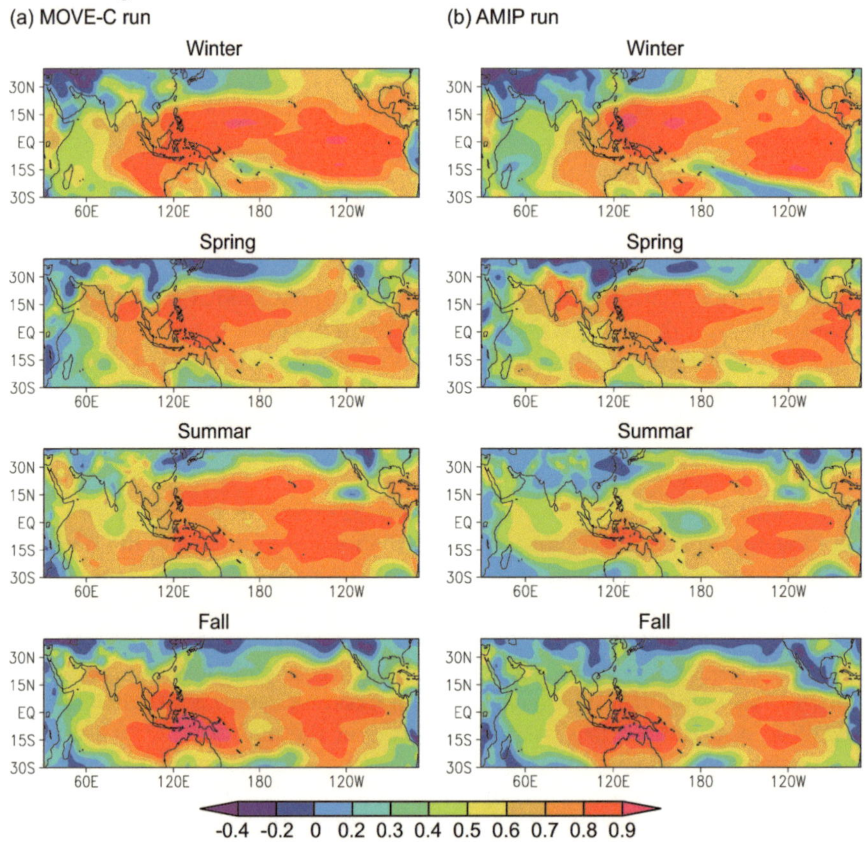

Fig. 8. ACC maps of ensemble mean SLP with the reference data in (a) MOVE-C run and (b) AMIP run for boreal winter, spring, summer, and fall. ACCs are calculated from the seasonal anomalies (deviation of the seasonal mean from the seasonal climatology) in 1980–2007.

In the MOVE-C run, the accuracy of SLP is also improved over that in the AMIP run. Figure 8 indicates the distribution of ACCs of SLP with the reference for boreal winter (December–February), spring (March–May), summer (June–August), and fall (September–November). The ACCs increase over the Indian Ocean for all seasons in the MOVE-C run. We also find that the MOVE-C run generally has higher ACCs around the central tropical Pacific. Furthermore, the ACCs over the Philippine Sea exceed 0.8 and are much higher than those in the AMIP run in summer.

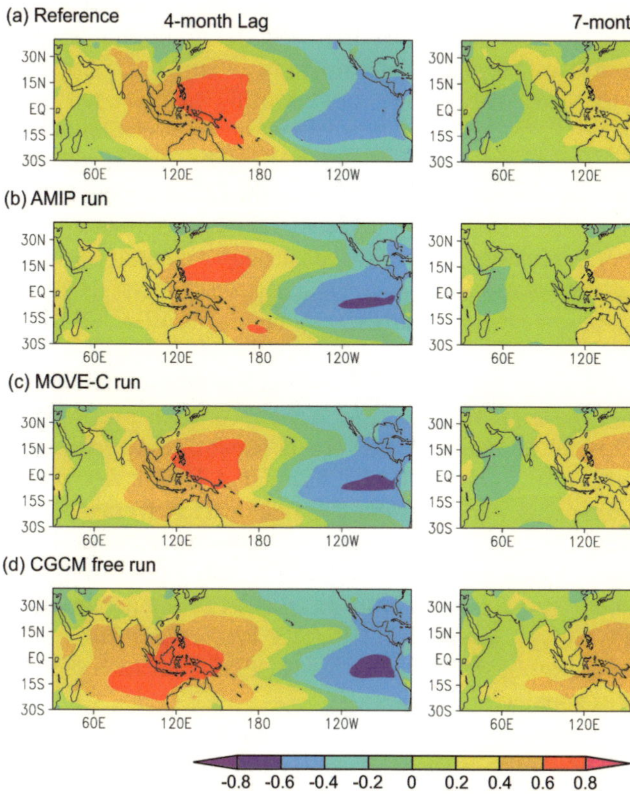

Fig. 9. Maps of the three-month and seven-month lagged correlation coefficients of the anomaly of SLP from its monthly climatology with the NINO3 index in (a) reference data, (b) AMIP run, (c) MOVE-C run, and (d) CGCM Free run. The coefficient is calculated for the NINO3 index in the period of 1980–2007. The anomaly is the deviation from the monthly climatology, and the three-month running mean is applied for both SLP and the index. The correlation coefficients averaged for ensemble members are adopted for the MOVE-C and AMIP runs.

These improvements are also due mainly to the better response of the atmospheric fields to ENSO. Figure 9 depicts maps of the four-month and seven-month lagged correlation of SLP with the NINO3 index. Since the peak of an El Niño is generally in December, the four-month and seven-month lagged correlations roughly represent the property of the spring and summer after El NIños.

The correlation coefficients are too high over the Indian Ocean in the CGCM free run (Fig. 9(d)). This spuriously high correlation may be due to insufficient representation of intrinsic variability associated with the Indian Ocean in the CGCM. In contrast, the correlation coefficients are underestimated over the Indian Ocean in the AMIP run (Fig. 9(c)). Furthermore, the AMIP run has lower correlation coefficients than the reference over the western equatorial Pacific and Philippine Sea. These errors are mostly removed in the MOVE-C run ((Fig. 9(b)), resulting in improved SLP accuracy shown in Fig. 8.

The better response of SLP to ENSO in the MOVE-C run is probably associated with improvement of the precipitation field over the South Indian Ocean, Philippine Sea, and western equatorial Pacific shown in Fig. 5. In Fujii et al. (2009), we indicated that the correlation between SST and precipitation is negative over the Philippine Sea and western equatorial Pacific, and around zero over the Indian Ocean in summer. This correlation is always spuriously high in the AMIP run, due to the lack of the negative feedback between SST and precipitation. This spurious correlation contaminates the variation of precipitation coupled to ENSO and reduces the correlation of SLP with ENSO in the AMIP run. This problem is mitigated by restoring the negative feedback by the CGCM in the MOVE-C run.

The previous study (Fujii et al., 2009) also suggested that appropriate reproduction of the variability of the Walker Circulation and monsoon trough is a factor which improves precipitation over the Philippine Sea in summer. In this study, we compare the response of the Walker Circulation and the monsoon trough to ENSO among the MOVE-C, AMIP, and CGCM free runs.

In order to represent the variations of the Walker Circulation and monsoon trough, we extend the W-Y index proposed by Webster & Yang (1992) and the DU2 index proposed by Wang & Fan (1999). The W-Y index is the anomaly of the vertical shear of the zonal wind between 850 hPa and 200 hPa averaged in 0–20°N, 40-120°E, and the DU2 index is the anomaly of the difference between the zonal winds on the 850 hPa surface averaged in the square areas of 5–15°N, 90-130°E and of 22.5-32.5°N, 110-140°E. A positive value of the W-Y (DU2) index denotes that the Walker Circulation (monsoon trough) is stronger than the usual. Both indices are originally calculated from the 3-month mean field of the zonal wind in boreal summer (June–August), and, thus, represents the state of the Walker Circulation and monsoon trough only in summer. In this study, these indices are calculated for each month using the three-month running mean of zonal wind fields.

Figure 10(a) demonstrates that the response of the Walker Circulation to ENSO is remarkably improved in the MOVE-C run over the other two runs. The plot of the reference data for the W-Y index indicates that the Walker Circulation is suppressed most around the peaks of the positive phases of ENSO (i.e., El Niño). The plot is almost symmetrical about the 0 lag line, and the minimum value reaches -0.7.

The response of the Walker Circulation is underestimated in the AMIP run, probably due to the weaker response of the velocity potential over the maritime continent (Fig. 6). This weak response mitigates the change of the zonal gradient of the velocity potential and reduces the change of the zonal wind in the upper layer, resulting in moderation of the change of the Walker Circulation. In contrast, the response of the Walker Circulation is overestimated and slightly lagged in the CGCM free run due to the stronger response of the velocity potential. In the MOVE-C run, the scale of the response is adequately estimated, and the plot is much closer to that for the reference, although the recovery of the Walker Circulation after the peaks is more rapid than the moderation before, as well as other two runs.

The response of the monsoon trough to ENSO is also reproduced best in the MOVE-C run as shown in Figure 10(b). The DU2 index is slightly negative four months before the peaks of El Niños, and it reaches the minimum value, about -0.5, six month after the peaks. This

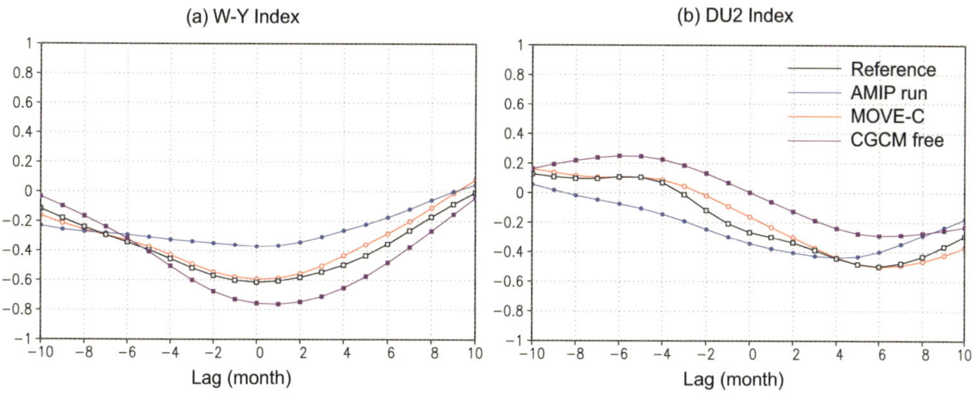

Fig. 10. Plots of the correlation coefficients of (a) W-Y index, (b) DU2 index, with the NINO3 index against the lag (month) of the W-Y or DU2 indices for the reference data (black), AMIP run (blue), MOVE-C run (red), and CGCM free run (purple). The coefficient is calculated for the NINO3 index in the period of 1980–2007. The correlation coefficients averaged for ensemble members are adopted for the MOVE-C and AMIP runs.

negative value represents the development of an anticyclone over the Philippine Sea after El Niños (eg., Wang et al., 2003; Xie et al., 2009). This property is recovered well in the MOVE-C run.

In contrast, in the AMIP run, no positive correlation before the peak is estimated, the lag of the minimum correlation is shorter than the real one, and the minimum value is slightly underestimated. The shorter lag and the weaker response are probably caused by the spuriously high correlation between SST and precipitation disturbing the atmospheric response, particularly in summer. The improved response of the DU2 index in the MOVE-C run over the AMIP run is associated with the better response of SLP over the Philippine Sea (Fig. 9) and the improved accuracy of SLP (Fig. 8). In the CGCM free run, the positive correlation before the peak of El NIños is overestimated and the negative correlation after the peaks is underestimated.

Finally, MOVE-C improves responses of the atmospheric circulation, including the Walker Circulation and the circulation associated with the monsoon trough, over the AMIP run, resulting in a better SLP and upper-tropospheric velocity potential field. This improvement stems from restoring the negative feedback between SST and precipitation in MOVE-C. The feedback adjusts the response of precipitation on the oceanic near-surface temperature field at an adequate level. Thus, calculating the coupled air-sea process explicitly through a CGCM (which is not possible with an AGCM alone) mitigates the inconsistency between the ocean and atmosphere and improves the representation of climate variability, including the ENSO response of the atmosphere. This result demonstrates a benefit of assimilating observation data directly into a CGCM in a coupled data assimilation system.

6. Summary

This chapter highlights the essential role of the ocean data assimilation systems for ENSO monitoring and SI forecasting. Considerable efforts to improve the data assimilation systems and increased ocean observation data have brought about the recent brilliant development of ocean data assimilation systems, contributing to the improvement of SI forecasting.

Although ENSO and seasonal forecasting have been realized and sophisticated enough to be used operationally now, further improvements are earnestly desired to reduce the damage of climate disasters and to increase the efficiency of industry, agriculture, and fisheries. Further development of ocean data assimilation systems is a possible factor for such improvement. Although the innovative progress of the development thus far has followed material changes of the ocean observing system as described in Section 2, no other material change is likely to occur in the next few years. Instead, we assume that imposing the ocean-atmosphere balance relationship to mitigate the coupled shock is the key to further improvement of the assimilated fields and SI forecasting, just as imposing the T-S balance relationship was the key to progress from the first generation to the second generation of ocean data assimilation systems.

In order to achieve such development, the use of a coupled atmosphere-ocean model is essential; thus, it is necessary to develop a system in which observation data are assimilated into a CGCM. In Section 5, we demonstrated a benefit of this strategy. However, the quasi-coupled data assimilation system introduced in section 5 is insufficient because it relies on adjustment in the coupled model integration for establishing atmosphere-ocean balance. Developing a scheme to assimilate atmospheric data is also desirable. Atmospheric data have an inherent potential to improve assimilated fields, as demonstrated in Balmaseda & Anderson (2009), although the difference in the major time-scale between the atmosphere and ocean makes the assimilation rather difficult.

To adress these issues, several institutes, including JMA/MRI, have currently been developing truly coupled data assimilation systems, in which oceanic and atmospheric data are assimilated into a coupled model imposing the ocean-atmosphere balance relationship. We expect those systems to form the third generation of data assimilation systems for ENSO monitoring and SI forecasting.

It should also be noted that a solid ocean observing system is essential for issuing reliable information on ENSO and seasonal forecasts. Therefore, sustaining current observing platforms, including the TAO/TRITON array and Argo floats, is crucially important, as well as proposing innovative and potential platforms. In order to sustain current platforms securely, it is necessary to demonstrate to the administrators the essential effects of those observation data in a readily visible manner as attempted in Section 4. In that sense, the activity of the GODAE Ocean View observing system evaluation task team (see the last of Section 4) should be seriously supported. In particular, we hope that the observation impact statement will have a substantial effect for the secure development of the ocean observing system, resulting in further improvements of ENSO monitoring and SI forecasting.

7. Acknowledgements

This study was partly supported by the Grant-in-Aids for Science Research 19540469, and 21540457 from the Ministry of Education, Culture, Sports, Science and Technology, Japan.

8. References

Ando, K. & McPhaden, M. J. (1997). Variability of surface layer hydrography in the tropical Pacific Ocean. *J. Geophys. Res.*, Vol. 102, 23,063–23,078.

Balmaseda, M. A. & Anderson, D. L. T. (2009). Impact of initialization strategies and observations on seasonal forecast skill. *Geophy. Res. Lett.*, Vol. 36, L01701, doi:10.1029/2008GL035561.

Balmaseda, M. A.; Alves, O. J.; Arribas, A.; Awaji, T.; Behringer, D. W.; Ferry, N.; Fujii, Y.; Lee, T.; Rienecker, M.; Rosati, T. & Stammer, D. (2009). Ocean initialization for seasonal forecasts. *Oceanogr.*, Vol. 22, No. 2, 154–159.

Balmaseda, M. A.; Dee, D.; Vialard, A. & Anderson, D. L. T. (2007). A multivariate treatment of bias for sequential data assimilation: application to the tropical oceans. *Q. J. R. Meteorol. Soc.*, Vol. 133, 167–179.

Balmaseda, M.; Fujii, Y.; Aves, O.; Awaji, T.; Behringer, D.; Ferry, N.; Lee, T.; Rienecker, M.; Rosati, T.; Stammer, D.; Smith, D. & Molteni, F. (2011). Initialization for Seasonal and Decadal Forecasts. *Proceedings of OceanObs'09: Sustained Ocean Observations and Information for Society*, Vol. 2, Hall, J.; Harroson, D. E. & Stammer, D., Eds., Venice, Italy, September 2009, ESA Publication WPP-306, doi:10.5270/OceanObs09.cwp.02.

Bloom, S. C.; Takacs, L. L.; Da Silva, A. M. & Ledvina, D. (1996). Data assimilation using incremental analysis updates. *Mon. Wea. Rev.*, Vol. 124, 1256–1271.

Carton, J. A.; Chepurin, G. & Cao, X. (2000). A simple ocean data assimilation analysis of the global upper ocean 1950-1995. Part I: Methodology. *J. Phys. Oceanogr.*, Vol. 30, 294–326.

Clark, C.; Harrison, D. E.; Johnson, M.; Ball, G.; Freeland, H.; Goni, G.; Hood, M.; McPhaden, M.; Meldrum, D.; Merrifield, M.; Roemmich, D.; Sabine, C.; Send, U.; Weller, R.; Wilson, S.; Benveniste, J.; Bonekamp, H.; Donlon, C.; Drinkwater, M.; Fellous, J.-L., Gohil, B. S.; Jacobs, G.; Le Traon, P.-Y.; Lindstrom, E.; Mingsen, L.; Nakagawa, K. & Parisot, F. (2009): An overview of global observing systems relevant to GODAE. *Oceanogr.*, Vol. 22, No. 2, 22–33.

Collecte Localisation, Satellites (CLS) (2004). SSALTO/DUACS user handbook: (M)SLA and (M)ADT near-real time and delayed time products, *CLS-DOS-NT-04*, Np. 103, CLS, Toulouse, France, 42pp.

Conkright, M. E.; Antonov, J. I.; Baranova, O.; Boyer T. P.; Garcia, H. E.; Gelfeld, R.; Johnson, D.; Locarnini, R. A.; O'Brien, Smolyar, I. & Stephens, C. (2002), World Ocean Database 2001, Vol. 1, introduction, Levitus, S. Ed., *NOAA Atlas NESDIS*, No. 42, U. S. Government Printing Office, Washington D. C., 167pp.

Cooper, N. S. (1988). The effect of salinity on tropical ocean models. *J. Phys. Oceanogr.*, Vol. 18, 697–707.

Dobricic, S.; Pinardi, N.; Dani, M.; Bonazzi, A.; Fratianni, C. & Tonani, M. (2005). Mediterranean forecasting system: An imporved assimilation scheme for sea-level anomaly and its validation. *Q. J. R. Meteorol. Soc.*, Vol. 131, 3627–3642.

Ezer, T. & Mellor, G. L. (1994). Continuous assimilation of Geosat altimeter data into tree-dimensional primitive equation Gulf Stream model. *J. Phys. Oceanogr.*, Vol. 24, 832–846.

Fujii, Y. (2005). Preconditioned Optimizing Utility for Large-dimensional analyses (POpULar). *J. Oceanogr.*, Vol. 61, 167–181.

Fujii, Y.; Ishizaki, S. & Kamachi, M. (2005). Application of nonlinear constraints in a three-dimensional variational ocean analysis. *J. Oceanogr.*, Vol. 61, 655–662.

Fujii, Y. & Kamachi, M. (2003). Three-dimensional analysis of temperature and salinity in the equatorial Pacific using a variational method with vertical coupled temperature-salinity empirical orthogonal function modes. *J. Geophys. Res.*, Vol. 108, No. C9, 3297, doi:10.1029/2002JC001745.

Fujii, Y.; Kamachi, M.; Matsumoto, S. & Ishiaki, S. (2012). *J. Climate*, Vol. 22, No. 20, 5541–5557. Barrier layer and relevant variability of the salinity field in the equatorial

Pacific estimated in an ocean reanalysis experiment. *Pure Appl. Geophys.*, Vol. 169, doi:10.1007/s00024-011-0387-y (in press).

Fujii, Y.; Matsumoto, S.; Kamachi, M. & Ishizaki, S. (2010). Estimation of the equatorial Pacific salinity field using ocean data assimilation system. *Adv. in Geosciences*, Vol. 18, 197–212.

Fujii, Y.; Nakaegawa, T.; Matsumoto, S.; Yasuda, T.; Yamanaka, G. & Kamachi, M. (2009). Coupled climate simulation by constraining ocean fields in a coupled model with ocean data. *J. Climate*, Vol. 22, No. 20, 5541–5557.

Fukumori, I. (2002). A partitioned Kalman filter and smoother. *Mon. Wea. Rev.*, Vol. 130, 1370–1383.

Gent, P. R. & McWilliams, J. C. (1990). Isopycnal mixing in ocean circulation models. *J. Phys. Oceanogr.*, Vol. 20, 150–155.

Haines, K.; Blower, J. D.; Drecourt, J.-P. & Liu, C. (2006). Salinity assimilation using S(T): covariance relationship. *Mon. Wea. Rev.*, Vol. 134, 759–771.

Hamilton, D. (1994). GTSPP builds an ocean temperature-salinity database, *Earth System Monitor*, Vol. 4, No. 4, 4–5.

Hayes, S. P.; Mangum, L. J.; Picaut, J.; A. Sumi, A. & Takeuchi, K. (1991). TOGA-TAO, a moored array for real-time measurements in tropical Pacific Ocean, *Bull. Amer. Meteor. Soc.*, Vol. 72, 339–347.

Huang, B.; Xue, Y. & Behringer, D. (2008). Impacts of Argo salinity in NCEP global ocean data assimilation system: the tropical Indian Ocean. *J. Geophys. Res.*, Vol. 113, C08002, doi:10.1029/2007JC004388.

Ishii, M.; Shouji, A.; Sugimoto, S. & Matsumoto, T. (2005). Objective analyses of sea-surface temperature and marine meteorological variables for the 20th century using ICOADS and the Kobe collection, *Intl. J. Climatol.*, Vol. 25, 865–879.

Ishizaki, H. & Motoi, T. (1999). Reevaluation of the Takano-Oonishi scheme for momentum advection on bottom relief in ocean models. *J. Atmos. Oceanic. Tech.*, Vol. 16, 1994–2010.

Ji, M.; Leetmaa, A. & Derber, J. (1995). An ocean analysis system for seasonal to interannual climate studies. *Mon. Wea. Rev.*, Vol. 123, 460–481.

Ji, M.; Reynolds, R. W. & Behringer, D. (2000). Use of TOPEX/Poseidon sea level data for ocean analysis and ENSO prediction. Some early results. *J. Climate*, Vol. 13, 216–231.

Kalnay, E.; Kanamitsu, M.; Kistler, R.; Collins, W.; Deaven, D.; Gandin L.; Iredell, M.; Saha, S.; White, G. & Woollen, J. (1996). The NCEP/NCAR 40-year reanalysis project. *Bull. Amer. Meteor. Soc.*, Vol. 77, 437–471.

Kamachi, M.; Kuragano, T.; Yoshioka, N.; Zhu, J. & Uboldi, F. (2001). Assimilation of satellite altimetry into a western north Pacific operational model. *Adv. in Atom. Sci.*, Vol. 18, 767–786.

Keppenne, C. L.; Rienecker, M. M.; Kurkowski, N. P. & Adamec, N. P. (2005). Ensemble Kalman filter assimilation of temperature and altimeter data with bias correction and application to seasonal prediction. *Nonlinear processes in Geophysics*, Vol. 12, 491–503.

Kuroda, Y. (2002). TRITON, present status and future plan, *Report for the international workshop for review of the tropical moored buoy network*, JAMSTEC, Yokosuka, Japan, 77pp.

Liebmann, B. & Smith, C. A. (1996). Description of a complete (interpolated) outgoing longwave radiation dataset. *Bull. Am. Meteorol. Soc.*, Vol. 77, 1275–1277.

Locarnini, R. A.; Mishonov, A. V.; Antonov, J. I.; Boyer, T. P.; Garcia, H. E.; Baranova, O. K.; Zweng, M. M. & Johnson, D. R. (2010). World Ocean Atlas 2009, Volume 1:

Temperature, Levitus, S., ed., *NOAA Atlas NESDIS*, No. 68, U.S. Government Printing Office, Washington, D.C., 184 pp.

Lukas, R. & Lindström, E. (1991). The mixed layer of the western equatorial Pacific ocean. *J. Geophys. Res.*, Vol. 96, suppl., 3343–3357.

Maes, C. (1998). Estimating the influence of salinity on sea level anomaly in the ocean. *Geophys. Res. Lett.*, Vol. 25, 3551–3554.

Maes, C.; Behringer, D.; Reynolds, R. W. & Ji, M. (2000). Retrospective analysis of the salinity variability in the western tropical Pacific Ocean using an indirect minimization approach. *J. Atmos. Oceanic Technol.*, Vol. 17, 512–524.

Maes, C. & Belamari, S. (2011). On the Impact of Salinity Barrier Layer on the Pacific Ocean Mean State and ENSO. *SOLA*, Vol. 7, 097–100, doi:10.2151/sola.2011-025.

Maes, C.; Ando, K.; Delcroix, T.; Kessler, W. S.; McPhaden, M. J. & Roemmich, D. (2006). Observed correlation of surface salinity, temperature and barrier layer at the eastern edge of the western Pacific warm pool. *Geophys. Res. Lett.*, Vol. 33, L06601, doi:10.1029/2005GL024772.

McPhaden, M. J.; Busalacchi, A. J.; Cheney, R.; Donguy, J.; Gage, K. S.; Halpern, D.; Ji, M.; Julian, P.; Meyers, G.; Mitchum, G. T.; Niiler, P. P.; Picaut, J.; Reynolds, R. W.; Smith, N. & Takeuchi, K. (1998). The tropical Ocean-Global Atmosphere observing system, a decade of progress. *J. Geophys. Res.*, Vol. 103, 14,169–14,240.

Noh, Y. & Kim, H. J. (1999). Simulations of temperature and turbulence structure of the oceanic boundary layer with the improved near-surface process. *J. Geophys. Res.*, Vol. 104, 15,621–15,634.

Oke, P. R.; Balmaseda, M. A.; Benkiran, M.; Cummings, J. A.; Dombrowsky, E.; Fujii, Y.; Guinehut, S.; Larnicol, G.; Le Traon, P.-Y. & Martin, M. J. (2009). Observing system evaluations using GODAE systems. *Oceanogr.*, Vol. 22, No. 2, 144–153.

Oke, P.; Balmaseda, M.; Benkiran, M.; Cummings, J.; Dombrowsky, E.; Fujii, Y.; Guinehut, S.; Larnicol, G.; Le Traon, P & Martin, M. (2011). Observational Requirements of GODAE Systems *Proceedings of OceanObs'09: Sustained Ocean Observations and Information for Society*, Vol. 2, Hall, J.; Harrison, D. E. & Stammer, D., Eds., Venice, Italy, September 2009, ESA Publication WPP-306, doi:10.5270/OceanObs09.cwp.67.

Onogi, K.; Tsutsui, J.; Koide, H.; Sakamoto, M.; Kobayashi, S.; Hatsushika, H.; Matsumoto, T.; Yamazaki, N.; Kamahori, H.; Takahashi, K.; Kadokura, S.; Wada, K.; Kato, K.; Oyama, R.; Ose, T.; Mannoji, N. & Taira, R. (2007). The JRA-25 Reanalysis, *J. Meteor. Soc. Japan*, Vol. 85, 369–432.

Picaut, J.; Masia, F.; & du Penhoat, Y. (1997). An advective-reflective conceptual model for ENSO. *Science*, Vol. 277, 663–666.

Ricci, S.; Weaver, T.; Vialard, J. & Rogel, P. (2005). Incorporating state-dependent temperature constraints in the background error covariance of variational ocean data assimilation. *Mon. Wea. Rev.*, Vol. 133, 317–328.

Sugiura, N.; Awaji, T.; Masuda, S.; Mochizuki, Y.; Toyoda, T.; Miyama, T.; Igarashi, H. & Ishikawa, Y. (2008): Development of a 4-dimensional variational coupled data assimilation system for enhanced analysis and prediction of seasonal to interannual climate variation. *J. Geohys. Res.*, Vol. 113, C10017, doi:10.1029/2008JC004741.

Redi, M. H. (1982). Oceanic isopycnal mixing by coordinate rotation. *J. Phys. Oceanogr.*, Vol. 12, 1154–1158.

Roemmich, D.; Morris, M.; Young, W. R. & Donguy, J.-R. (1994). Fresh equatorial jet. *J. Phys. Oceanogr.*, Vol. 24, 540–558.

Takaya, Y.; Yasuda, T.; Ose, T. & Nakaegawa, T. (2010). Predictability of the Mean Location of Typhoon Formation in a Seasonal Prediction Experiment with a Coupled General Circulation Model. *J. Meteor. Soc. Japan*, Vol. 88, 799–812, doi:10.2151/jmsj.2010-502.

Troccoli, A.; Balmaseda, M. A.; Segschneider, J.; Vialard, J.; Anderson, D. L. T.; Haines, K.; Stockdale, T. N.; Vitart, F. & Fox, A. D. (2002). Salinity adjustments in the presence of temperature data assimilation. *Mon. Wea. Rev.*, Vol. 130, 89–102.

Tsujino, H.; Motoi, T.; Ishikawa, I.; Hirabara, M.; Nakano, H.; Yamanaka, G. & Yasuda, T. (2010). Meteorological Research Institute Community Ocean Model Version 3 (MRI.COM3) Manual. *Technical Repprt of Meteorological Research Institute*, No. 59, Meteorological Research Institute, Tsukuba, Japan, 241pp.

Usui, N.; Ishizaki S.; Fujii Y.; Tsujino, H.; Yasuda, T. & Kamachi, M. (2006). Meteorological Research Institute multivariate ocean variational estimation (MOVE) system: Some early results. *Adv. Space Res.*, Vol. 37, 806–822, doi:10.1016/j.asr.2005.09.022.

Vialard, J.; Delecluse, P. & Menkes, C. (2002). A modeling study of salinity variability and its effect in the tropical Pacific Ocean during the 1993-1999 period. *J. Geophys. Res.*, Vol. 107, No. C12, 8005, doi:10.1029/2000JC000758.

Wang, B. & Fan, Z. (1999). Choice of South Asian summer monsoon indices. *Bull. Am. Meteor. Soc.*, Vol. 80, 629–638.

Wang, B.; Wu, R. & Li, T. (2003). Atmosphere - warm ocean interaction and its impacts on Asian-Australian monsoon variation. *J. Climate*, Vol. 16, 1195–1211.

Weaver, A. T.; Vialard, J. & Anderson, D. L. T. (2003). Three- and four-dimensional variational assimilation with a general circulation model of the tropical Pacific Ocean. Part I: Formulation, internal diagnostics, and consistency, checks. *Mon. Wea. Rev.*, Vol. 131, 1360–1378.

Webster, P. J. & Yang, S. (1992). Monsoon and ENSO: selectively interactive systems. *Q. J. R. Meteorol. Soc.*, Vol. 118, 877–926.

Woodgate, R. A. (1997). Can we assimilate temperature data alone into a full equation of state model? *Ocean Modeling* (unpublished manuscripts), Vol. 114, 4–5.

Xie, S. P.; Hu, K.; Hafner, J.; Tokinaga, H.; Du, Y.; Huang, G. & Sampe, T. (2009). Indian ocean capacitor effect on Indo-Western Pacific climate during the summer following El Niño. *J. Climate*, Vol. 22, 730–747.

Yan, C.; Zhu, J.; Li, R. & G. Zhou, G. (2004). Roles of vertical correlations of background error and T-S relations in estimation of temperature and salinity profiles from sea surface dynamic height. *J. Geophys. Res.*, Vol. 109, C08010, doi:10.1029/2003JC002244.

Yasuda, T.; Takaya, Y.; Kobayashi, C.; Kamachi, M.; Kamahori, H. & Ose, T. (2007). Asian monsoon predictability in JMA/MRI seasonal forecast system. *CLIVAR Exchanges*, Vol. 43, 18–24.

Zhang, S.; Harrison, M. J.; Rosati, A. & Wittenberg, A. (2007). System design and evaluation of coupled ensemble data assimilation for global oceanic studies. *Mon. Wea. Rev.*, Vol. 135, 3,541–3,564.

5

ENSO-Type Wind Stress Field Influence over Global Ocean Volume and Heat Transports

Luiz Paulo de Freitas Assad
Federal University of Rio de Janeiro/COPPE/LAMCE
Brazil

1. Introduction

Ocean-atmosphere interactions have a fundamentally important effect on the heat energy provided by the sun. Solar radiation that reaches Earth's surface generates heat gain in tropical ocean regions and heat loss in high latitudes. This energy is redistributed above the continents exclusively via the atmosphere, whereas energy redistribution above the oceans occurs through a cooperative relationship between the oceans and the atmosphere (Hastenrath, 1979). Knowledge of the relationships between oceanic and atmospheric processes and their variability are of great importance to Earth climate studies. Oceanic and atmospheric climate variability may be associated with interactive processes, such as the well-studied El Niño Southern Oscillation (ENSO). Although ENSO processes are centred in the Equatorial Pacific region, changes in tropical atmospheric convection could affect the atmospheric circulation of the entire planet. This perturbed atmospheric circulation, particularly the low-level wind stress field, could affect not only ocean transport but also the spatial and temporal distribution of certain dynamic and thermodynamic ocean properties, especially in the upper layers of the ocean (Colberg and Reason, 2004).

In some regions, such as the Antarctic Circumpolar Current (ACC) system, most of the variability in volume transport is associated with the barotropic field. High correlation values between the deep ocean pressure field and the wind stress field can be observed at the southern side of the Drake Passage (Whitworth & Peterson, 1985). Interannual variability in ACC volume transport has also been associated with global climate variability phenomena such as ENSO. Lenn et al. (2007), analysing five years (1999-2004) of ADCP (High-resolution Acoustic Doppler-Current Profiler) datasets from the Drake Passage from three repeated cross channel tracks, found negative surface layer (< 250 meters) volume transport anomalies during the period between 2002 and 2003, which were ENSO years. Even though this result is inconclusive, it suggests a weakening of the surface ACC flow in the Drake Passage associated with the 2002-2003 ENSO event.

The influences of global interannual variability are not only observed in large-scale ocean features. Lentini et al. (2001) studied sea surface temperature variability in the southwestern Atlantic region. The authors observed a southward and northward advection of cold and warm anomalies (±1°C) during and immediately after ENSO events and noted that the larger amplitudes of these anomalies were situated at the La Plata River and Patos Lagoon offshore regions. Assad et al. (2010) studied the influence of an ENSO type global wind stress field on the upper ocean temperature anomalies over the Brazil - Malvinas

Confluence region and observed positive and negative temperature anomalies associated with upper layer volume transport changes in the Brazil Current. The authors also observed changes in the upper layer of ACC volume transport.

This work aims to identify and quantify the occurrence and influence of global ENSO-type wind stress forces on global ocean heat and volume transports and their interannual climate variability using an Ocean General Circulation Model (OGCM). This phenomenon could cause climate changes in the amount of heat storage in each ocean basin, which could induce climate variability in the nearby continental regions. This kind of information would be very useful to be applied in global climate models in order to better understand how heat energy would be distributed over the planet by the occurrence of an ENSO event.

The ocean model was forced with a three-year transient ENSO-type global wind stress field. The results show significant changes in the global ocean integrated kinetic energy field that represent a direct response to changes in the volume transports integrated for different density water levels around the world's ocean basins. Changes in the ocean volume transport around the world were observed not only in the upper density layers of the ocean but also, to a minor degree, in the deep and bottom density layers. These observed changes in ocean volume transports caused changes in the global water column integrated heat transport, which, in turn, caused significant changes in the global meridional overturning cell circulation along the perturbed wind experiment.

2. Methodology

This work applied an Ocean General Circulation Model (OGCM) as a tool for investigating the variability of ocean volume and heat transports in the world's oceans. The Modular Ocean Model 4.0 (Pacanowsky & Griffies, 1999), developed by the Geophysical Fluid Dynamics Laboratory (GFDL, National Oceanic and Atmospheric Administration, Princeton, New Jersey, USA), was used for this study. This verical z-coordinate model has been largely adopted by the scientific community for global climatic research investigations. It is mathematically formulated with a Navier-Stokes set of equations under the Boussinesq and hydrostatic approximations. The system of continuous equations are basically completed by an equation of state of sea-water, the continuity equation for an incompressible fluid, and conservation equations for temperature and salinity.

The longitudinal resolution of the model was 1° worldwide and the latitudinal resolution increased from 1° to 1/3° within the 10° N-10° S equatorial band. Fifty levels were implemented on the vertical axis. To achieve a higher resolution near the ocean surface, the first 22 levels were located in the top 220 m of the water column. The thickness of each level varied from 10 m to a maximum of 366.6 m near the sea bottom (5500 m). Regions shallower than 40 m were not considered. The grid constructed on this work uses the tripolar grid method by Murray (1996). It adopts normal spherical coordinates south of 65° N while North of this latitude two numerical poles situated on land regions of Asia and North America are used. This method minimizes problems related with the convergence of meridians over the ocean that happens mainly in North Pole.This grid followed the oceanic component of the GFDL climate model presented in the 4th IPCC Assessment of Global Climate Changes. Boundary conditions of the OGCM were established by fields of wind stress, sea surface temperature, sea surface salinity, sea surface heat fluxes, sea surface elevation, runoff and precipitation. The model used three-dimensional temperature, salinity and velocity as the initial conditions. All of boundary and initial conditions will be further detailed in the next section.

Two experiments were conducted to estimate anomalous ocean volume and heat transports. The first sought to obtain the ocean climate transports and was used as a control parameter to determine ocean transport anomalies during ENSO conditions. The second experiment constituted a perturbed condition, where an ENSO-type wind stress field was imposed as a boundary condition. These anomalous boundary conditions were represented by monthly mean global wind stress fields generated by an AGCM experiment (Torres Jr., 2005). The AGCM was integrated for three years and used boundary conditions based on the 1982–1983 ENSO event.

The ocean volume and heat transport were estimated for selected zonal and meridional sections of the entire world ocean (Figure 1). The choice of these sections followed two basic criteria: the existence of estimates in the scientific literature and the dynamic relevance of the analysed region in the global ocean circulation. Ocean volume transports were estimated for three different density (σ) levels in units of kg.m^{-3}: upper ($\sigma < 27.5$), deep ($27.5 < \sigma < 27.8$) and bottom ($\sigma > 27.8$). These three levels were chosen based on the study conducted by Schmitz (1996). The author used data sets from the World Ocean Circulation Experiment to study the global ocean circulation transports in these three different levels. In fact, the upper layer chosen for the current study is a compilation of surface and intermediate layers investigated by Schmitz (1996) as follows: Surface Layer Water, Upper Layer Water, Sub-Antarctic Mode Water (SAMW), upper SAMW, lower SAMW and total intermediate water (IW). The ocean heat transports were estimated for the whole water column in each monitored section.

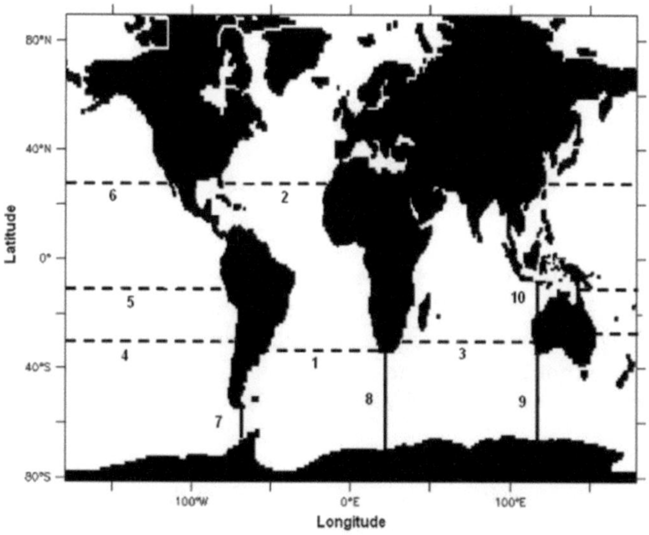

Sections	Position
1	32⁰ South Atlantic
2	25⁰ North Atlantic
3	30⁰ South Indian
4	30⁰ South Pacific
5	10⁰ South Pacific
6	25⁰ North Pacific
7	Drake Passage (70⁰ W)
8	Africa / Antartica (20⁰ E)
9	Australia / Antartica (120⁰ E)
10	Indonesia

Fig. 1. Sections where volume and heat transports were estimated. Each section is identified by a number and a name.

The geometry of each monitored section together with the annual climatological mean space distributions of the analysed density levels are illustrated in Figure 2. The annual mean bottom layer density levels were not obtained in only two sections: Indonesia and 25⁰ North Pacific. It is also important to note the presence of known density distribution behaviours, as the strong eastward low level winds around the Antarctic continent generate ocean current via Ekman transport and induce the sloping of isopycnals observed in ACC meridional sections (Schmitz, 1996; Russel, 2006). In the other zonal sections the isopycnal lines are not as sloped as meridional sections reflecting more water column stratification, observed in these regions (Schmitz, 1996).

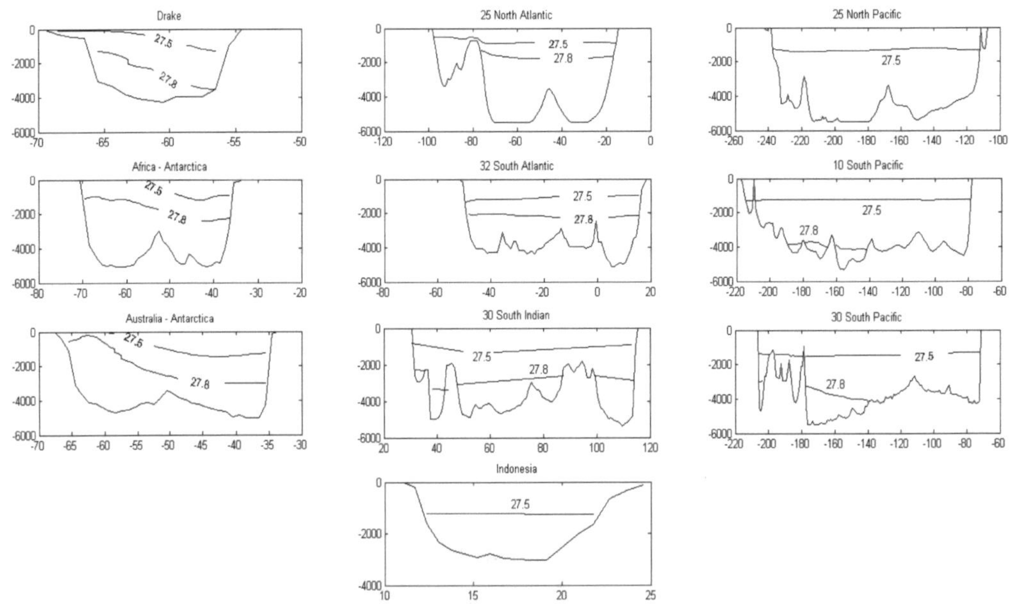

Fig. 2. Geometry of the monitored sections with 27.5 and 27.8 control experiment isopycnal lines (in units of kg.m^{-3}) superimposed.

2.1 Control experiment

As previously described, this experiment mainly aims to estimate the climate behaviour of volume and heat global ocean transports. The three-dimensional temperature, salinity and velocity data sets from the Ocean Data Assimilation for Seasonal to Interannual Prediction experiment (ODASI) conducted by GFDL were used as the initial condition in the climate experiment (Sun et al., 2007). The month of January 1985 was chosen to be the initial condition for the climate experiment because it did not present strong climate variability phenomena such as El Niño.

The sea surface boundary conditions were taken from the climatological data set of the Ocean Model Intercomparison Project (OMIP). This version of the OMIP data set was produced by the ECMWF (European Centre for Medium-Range Weather Forecasts) under project ERA-15 (Röeske, 2001). The variables used for surface forcing were meridional and zonal wind stress components, net short- and long-wave radiation, sensible heat flux, specific humidity flux, frozen and liquid precipitation and river runoff. The boundary fields given above were cyclically imposed on the model until it reached an equilibrium state. The model was integrated for seven years with an external mode time step of 80 s and a baroclinic time-step of 4800 s. Using 8 processors of an ALTIX 350 computer, the model was able to complete one simulated year in approximately 1.5 days, generating approximately 2 GB output per model year.

2.2 Perturbed experiment

The main goal of this experiment was to identify interannual variability in the global ocean transports induced by a three-year transient wind stress field. The perturbed experiment used the September monthly mean dynamic and thermodynamic fields generated by the control experiment as initial conditions, but with only anomalous wind stress boundary conditions. These anomalous boundary conditions were determined by monthly mean global wind stress fields generated by an atmospheric general circulation model experiment (AGCM). The AGCM was integrated for three years, and it used Pacific Equatorial sea surface temperature boundary conditions based on the 1982–1983 ENSO event (Torres Junior, 2005; Assad et al., 2010). The ocean model was integrated for 36 months (three years) in this experiment. The annual mean values for volume and heat transports in each monitored section were then estimated and analysed.

3. Results

3.1 Control experiment

The integrated global ocean kinetic energy shows a well-defined cyclic behaviour starting at the second integration year (Fig. 3). The most energetic periods occur during austral winter and the less energetic periods occur during austral summer. This result indicates that the ocean kinetic energy annual cycle is dominated by the wind stress annual energy cycle (Oort et al., 1994).

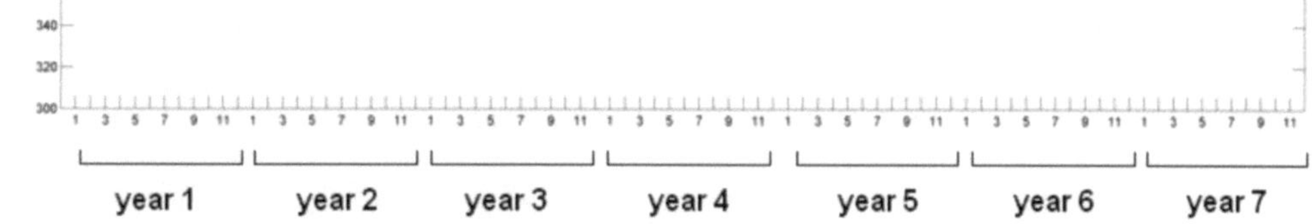

Fig. 3. Time series of the integrated ocean kinetic energy for the whole integration domain in PW (PW = 10^{15} W).

3.1.1 Volume transport

The obtained volume transports showed good agreement with the published values. The values obtained here also represent important aspects of wind-driven ocean climates and thermohaline circulation. Figure 4 shows the directions and intensities (units of Sv) of the annual mean ocean volume transports that were vertically integrated for the three different density levels for each monitored section.

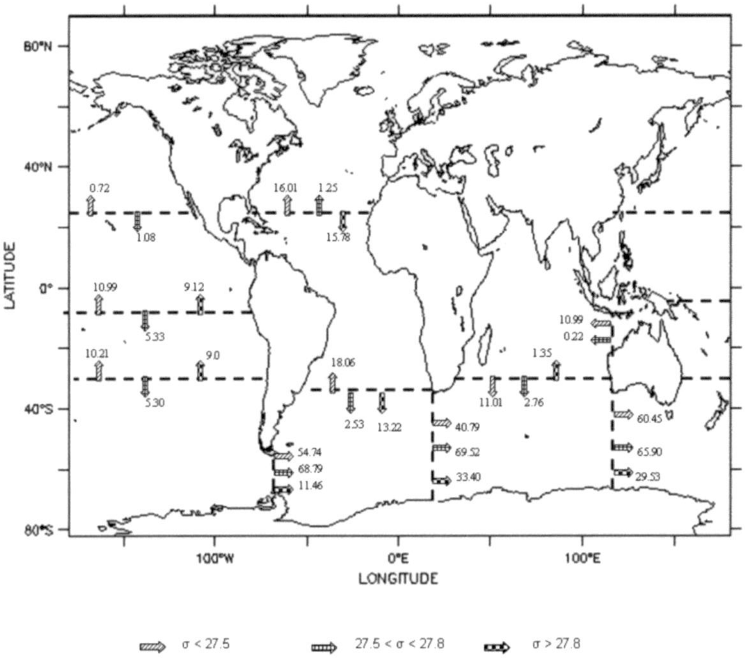

Fig. 4. Annual mean ocean volume transports for the defined density levels.

The analysis of this map revealed, for example, a strong eastward flow for the three monitored density levels associated with the ACC around the Antarctic continent. The obtained vertically integrated volume transport in this section is 134.99 Sv. Ganachaud & Wunsch (2000), using hydrographic data from the World Ocean Circulation Experiment (WOCE), estimated a value of 140 Sv (± 6) for a section situated near the Drake Passage. Stammer et al. (2003), using an OGCM constrained by WOCE project data, obtained a value of 134 Sv (± 7) for a Drake Passage section. The model was also able to reproduce the seasonal cycle of ACC volume transport, with higher values during austral winter months and lower values during austral summer months (Figure 5).

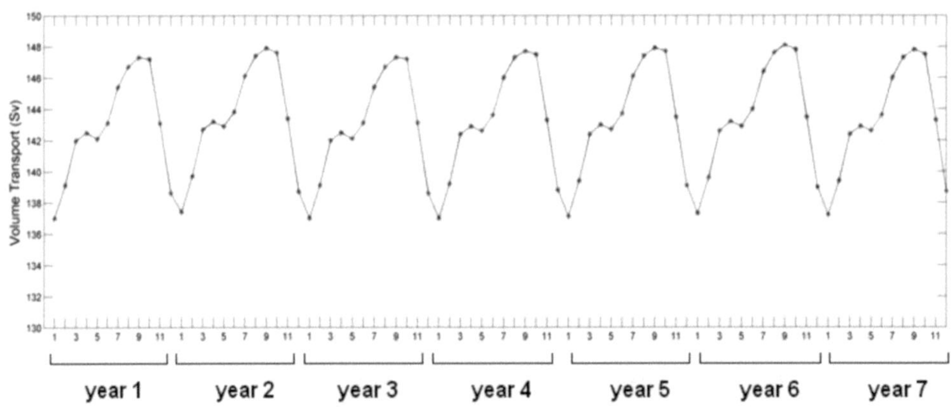

Fig. 5. Time series of zonal volume transport across the Drake Passage (Fig. 2; section 7) in Sv.

For a South Atlantic zonal section situated near 25^0 S, Ganachaud & Wunsch (2000) estimated a value of 16 Sv (± 3) for an upper level layer northward transport with $\sigma < 27.72$. The authors also estimated a resultant southward transport of 17 Sv for a South Atlantic zonal section situated near 25^0 S for a layer with $\sigma > 27.72$. For the 32^0 South Atlantic section monitored in the present study, a value of 18.06 Sv was obtained for the northward upper density level layer. For the bottom layer, a southward volume transport of 13.22 Sv was obtained.

The 25^0 North Atlantic section revealed a 16.01 Sv northward volume transport for the upper layer. The resultant integrated volume transport of deep and bottom layers was in the southward direction, with an intensity of 14.53 Sv. Ganachaud & Wunsch (2000) estimated a southward volume transport with an intensity of 17 Sv in a section situated near the 25^0 North Atlantic section.

The westward volume transport of the Indonesian Throughflow was also observed for the upper and deep layers, and it showed a total transport of 11.21 Sv (Figure 2). Ganachaud & Wunsch (2000) estimated a value of 16 Sv (± 5) for the total westward transport. Stammer et al. (2003) obtained a value of 11.5 Sv (± 5) for this section.

In the 30^0 S and 10^0 S South Pacific sections, a northward flow in the upper and bottom layers was observed, whereas a southward flow was observed in the intermediate layer. The volume transport of the upper layer values in these sections were 10.21 Sv and 10.99 Sv, respectively. For a Pacific zonal section situated near 20^0 S, Ganachaud & Wunsch (2000) estimated a northward volume transport of 19 Sv (± 5) in an upper layer with $\sigma < 27.2$ (Stammer et al., 2003)

3.1.2 Heat transport

For the purposes of this study, the term "heat transport" refers specifically to heat advection (advective heat transport) associated with the mean ocean flow. The vertically integrated mean (from the sea surface to sea bottom) heat transport and respective variance across each monitored section are plotted in Figure 6. The Atlantic and Pacific zonal sections showed northward heat transports, whereas the Indian Ocean zonal section presented a southward heat transport (Stammer et al., 2003; Ganachaud & Wunsch, 2000). It is important to observe the significant variance values mainly for the 10^0 South Pacific section. The South Atlantic section exhibited a northward heat transport of 0.60 ± 0.02 PW (1 PW = 10^{15} Watts), reinforcing the peculiar behaviour of the South Atlantic basin as an exporter of heat to the North Atlantic. Using hydrographic data, Rintoul (1991) estimated the average value of 0.25 PW (± 0.12) for a zonal section in the South Atlantic at 32° S. Hastenrath (1980, *apud* Rintoul, 1991) estimated a value of 1.15 PW for the meridional heat transport in the South Atlantic basin.

The most intense heat transports are associated with the ACC zonal flux through its meridional sections (Fig. 1; sections 7, 8, and 9). The time series plot for the vertically integrated heat transport perpendicular to each monitored section also reveals a well defined annual cyclic behaviour (not shown). For the Drake Passage section (Fig. 1; section 7), the maximum values are found in austral summer and the minimum values in austral winter. The magnitude of this transport ranges from 1.48 to 1.58 PW with a mean value of 1.52 PW. These values agree with the previous estimate of 1.3 PW (Ganachaud & Wunsch 2000) and 1.14 PW (±0.06) obtained by Stammer et al. (2003). A less intense heat transport was observed in the Africa-Antarctica section (0.98 PW), which could be explained by the influence of the westward Agulhas Current flow, which is present in the upper layers of this section.

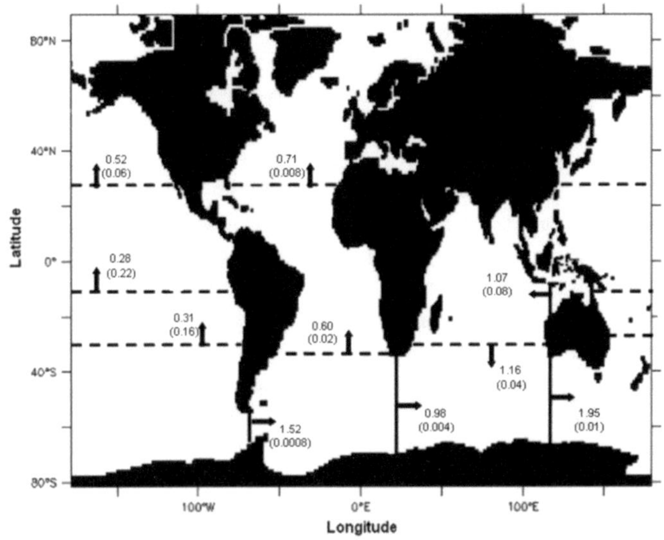

Fig. 6. Vertically integrated annual mean heat transport (direction and intensity) and associated variance (in PW^2) in each monitored section for the entire water column.

Figure 7 presents the latitudinal distribution of the whole water column integrated meridional heat transport over the world's ocean basins. This figure reveals the presence of an intense northward heat transport between the equator and 40° N and a southward heat transport between the equator and 40° S, which indicates that the conducted experiment was able to reproduce the so-called conveyor belt circulation. Using a geostrophic method on data obtained during the WOCE project, Ganachaud & Wunsch (2000) estimated a mean value of -0.6 PW (± 0.3) for the global meridional heat transport across 30° S and 2.2 PW (± 0.6) across 12° N. At the same latitudes, we found values of -0.2 PW (± 0.01) and 1.75 (with variance of 0.03 PW^2), respectively.

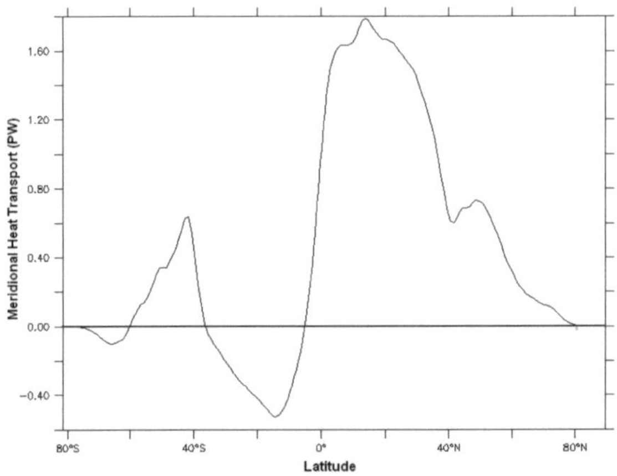

Fig. 7. Global integrated annual mean meridional heat transport (in PW) for the control experiment.

3.2 ENSO perturbation experiment

The global ocean integrated kinetic energy anomaly time series during the perturbed experiment is shown in Figure 8. This figure reveals the transient behaviour of this experiment, where the global ocean kinetic energy curve follows the global integrated work performed by the wind curve with a small time lag. The second year presented the most energetic months, and the global ocean kinetic energy maximum peak occurred in the 23rd month of integration. The global ocean work performed by the wind stress field over the sea surface had an energy peak in the 21st month. This result suggests a time lag of approximately two months between the peaks in global wind energy that led to the global ocean kinetic energy. It is important to emphasise that ocean dynamic and thermodynamic anomalies were observed before and after the ocean kinetic energy peak that occurred during the second integration year.

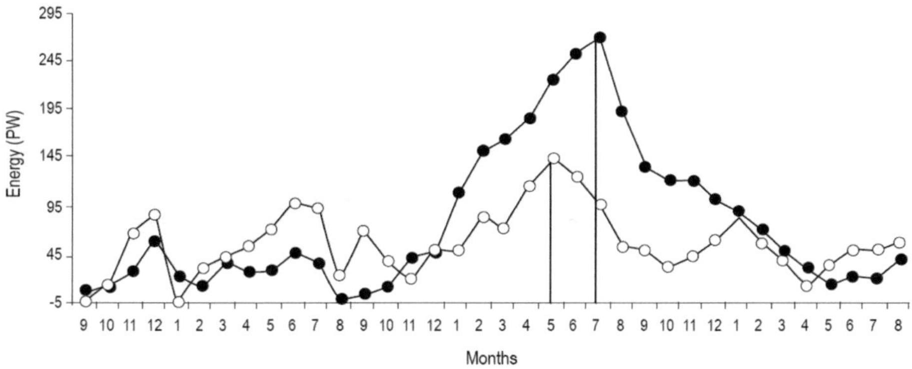

Fig. 8. Kinetic energy time series plot (filled circles for the ocean kinetic energy and empty circles for the global wind energy over the sea surface). The values of the work performed by the wind were multiplied by a factor of 10^3.

The annual mean global wind stress field for the control experiment and the anomaly wind stress field for each perturbed simulated year are presented in Figure 9. This figure reveals the presence of a significant anomalous wind stress circulation in all the world's ocean basins over the course of the simulated three year experiment. Specifically, there were obvious differences in the westward wind stress vectors around the Antarctic continent. These features reveal a decrease in the eastward low level wind circulation in this region throughout the perturbed experiment. It is also important to note the increased presence of anomalous wind stress difference vectors in the low and mid latitude ocean regions during the second integration year. These regions exhibited lower values in the first and third years, indicating less intense changes in the wind stress field during these years. The North Atlantic basin exhibited a decrease in the low level atmospheric subtropical gyre for the first year and an increase for the second perturbed year. The South Atlantic basin also exhibited a small increase of the low level atmospheric subtropical gyre during the second integration year. The Equatorial Pacific region exhibited the most intense differences in wind stress vectors fields during the second integration year, and the southeast trade winds system seemed to increase over the Pacific Ocean region this same year.

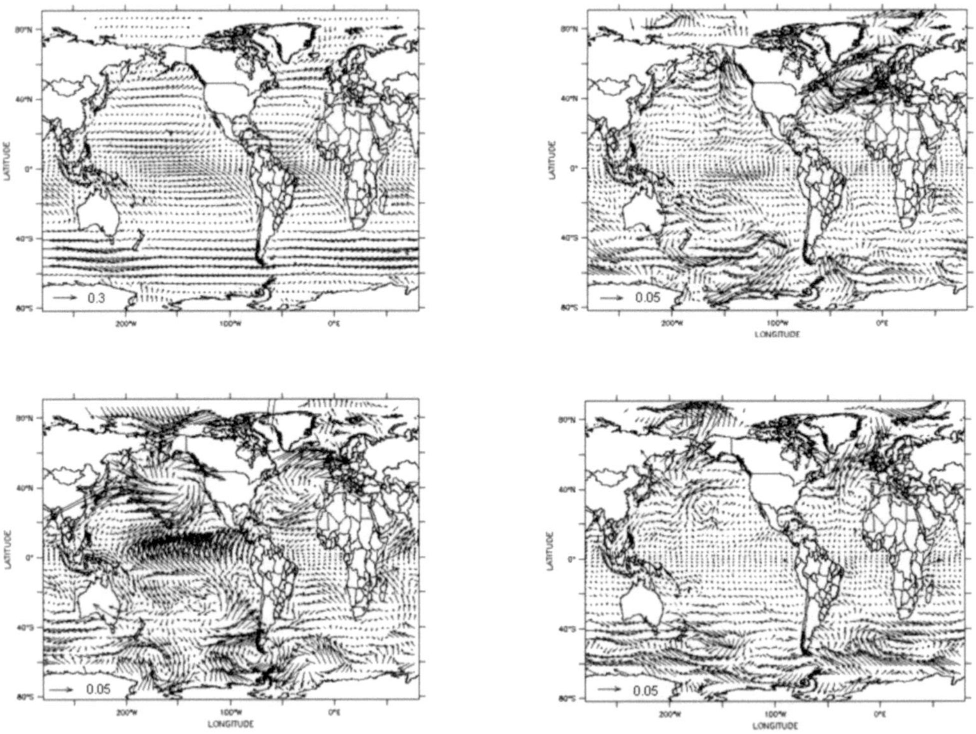

Fig. 9. Annual mean wind stress climatological vector field (upper left). Annual mean anomaly wind stress field for years 1 (upper right), 2 (lower left) and 3 (lower right). All values are in Pa.

3.2.1 Volume Transport

The annual mean volume transports integrated for the three selected levels in each monitored section for the first integration year are represented in Figure 10. This figure reveals an intensification of the northward upper layer volume transport for both Atlantic zonal sections (32^{0} South Atlantic and 25^{0} North Atlantic). For the bottom layer, 32^{0} South Atlantic exhibited an increase of southward volume transport, and the 25^{0} North Atlantic section exhibited a small decrease of the southward volume transport. The ACC meridional sections of the Drake Passage and Australia-Antarctica exhibited an increase of the eastward upper layer volume transport, while the Africa-Antarctica section exhibited a small decrease. The deep layer volume transport decreased for all ACC sections. The bottom layer showed an intensification of the eastward volume transport for the two ACC zonal sections bordering the South Atlantic basin. The Indonesian Throughflow section exhibited a small decrease of the upper layer volume transport for this year. The 30^{0} South Indian section exhibited an increase in the upper and bottom layer volume transports, while a decrease in the deep layer was observed. The 30^{0} and 10^{0} South Pacific sections exhibited a decrease in the northward upper layer volume transport, while an increase was observed in the 25^{0} North Atlantic upper layer volume transport. The bottom layer volume transport of the Pacific sections decreased in a volume transport decrease relative to control experiment values.

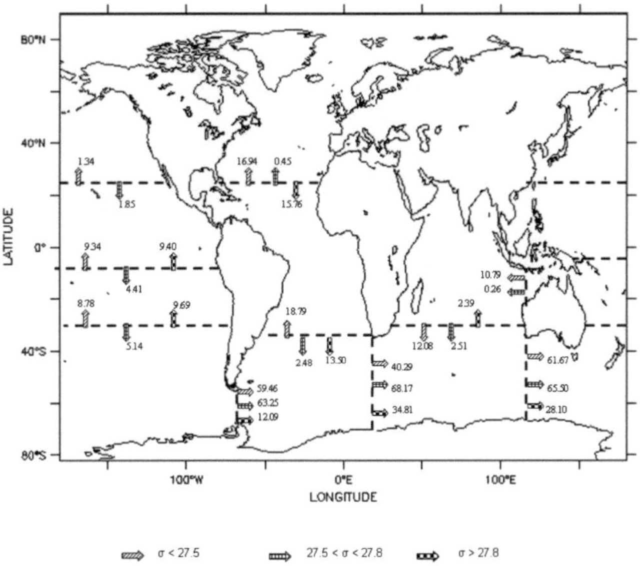

Fig. 10. Annual mean ocean volume transports for the selected levels in each monitored section for each defined density level in the first perturbed integration year.

The annual mean volume transports integrated for the three selected levels in each monitored section for the second integration year are represented in Figure 11. As explained in Figure 8, this period was the most energetic year. As in the first integration year, positive and negative anomalies were observed throughout all the world's ocean basins. The 32^0 South Atlantic section volume transports decreased in all the analysed density layers. The 25^0 North Atlantic section exhibited intensification of the northward upper layer transport and in southward bottom layer transport. The deep layer volume transport decreased and changed direction toward the south during the second integration year. Almost all the ACC meridional sections showed a decrease in volume transport during the second integration year. The intermediate density level volume transport in the Drake Passage section was the only exception, which increased by 1.65 Sv during this year. The 30^0 South Pacific and 25^0 North Pacific zonal sections exhibited an intensification of the upper level northward volume transport, while the 10^0 South Pacific section exhibited a decrease in the upper layer northward volume transport. The deep layer southward volume transport increased in all three Pacific zonal sections. The bottom layer transports decreased. The Indonesian Through flow eastward volume transport decreased during the second integration year. The 30^0 South Indian upper and bottom layer volume transports exhibited a southward and northward intensification, respectively.

The annual mean volume transports integrated for the three selected levels in each monitored section for the third integration was characterised by a decrease of the global integrated ocean kinetic energy in relation to the second integration year (Figure 8). The 32^0 South Atlantic section exhibited the most intense northward upper layer volume transport value. The southward bottom layer transport for this section also intensified slightly, while the southward deep layer transport exhibited a small decrease. The 25^0 North Atlantic section exhibited an intensification of the northward upper layer volume transport, while a decrease was observed in the deep layer southward transport. For the deep layer transport,

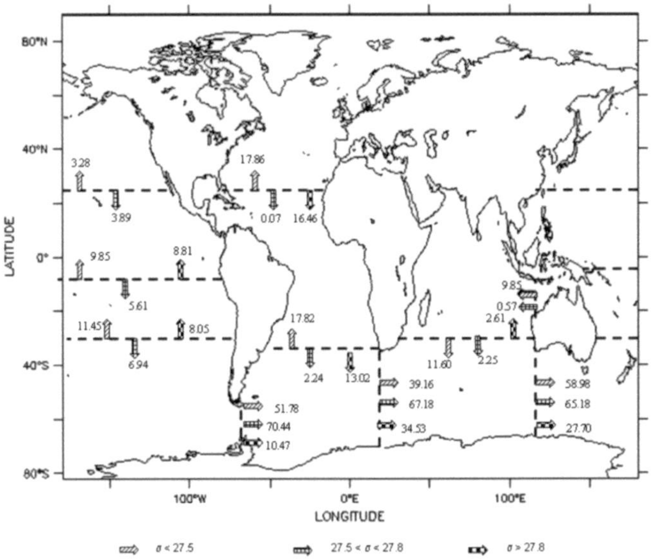

Fig. 11. Annual mean ocean volume transports for the selected levels in each monitored section for each defined density level in the second perturbed integration year.

an intensification of the southward volume transport in relation to the value observed in the second integration year is apparent. All the ACC upper layer volume transport values exhibited a decrease in their intensity compared with the control experiment values. The upper and deep layer volume transports of the Pacific sections intensified. The bottom layer transports decreased in all monitored Pacific sections. The Indonesian Throughflow westward volume transport exhibited the lowest values of the three integration perturbed years. The 30^0 South Indian volume transports decreased in all the monitored density layers.

3.2.2 Heat transport

The global meridional heat transport integrated for the entire water column showed significant changes for all simulated perturbed years. Figure 12 presents the integrated heat transport across each monitored section for the first perturbed year. The 32^0 South Atlantic section demonstrated a small increase of the northward heat transport (0.64 PW) compared with the control experiment value (0.60 PW). The 25^0 North Atlantic section also showed an increase in the northward heat transport, at 1.10 PW. The three ACC sections exhibited an intensification of the eastward heat transport. The increase in the northward meridional heat transport for the Atlantic sections could be related to the upper layer volume transport intensification of both sections as observed in Figure 10 The Indonesian Through flow section exhibited a decrease in the westward heat transport (0.82 PW) compared to that of the control experiment (1.07 PW). In the Pacific basin, a decrease in the northward heat transports of the 30^0 South Pacific and 10^0 South Pacific sections was observed, whereas there was an increase in the northward heat transport across the 25^0 North Pacific section. These results can be explained by the data illustrated in Figure 10, which reveals an increase in the upper layer volume transport of the two South Pacific sections, and a decrease in volume transport in the upper layer at 25^0 North Pacific. The 30^0 South Indian section exhibited an increase in the southward heat transport. A value of 1.29 PW was observed for this section.

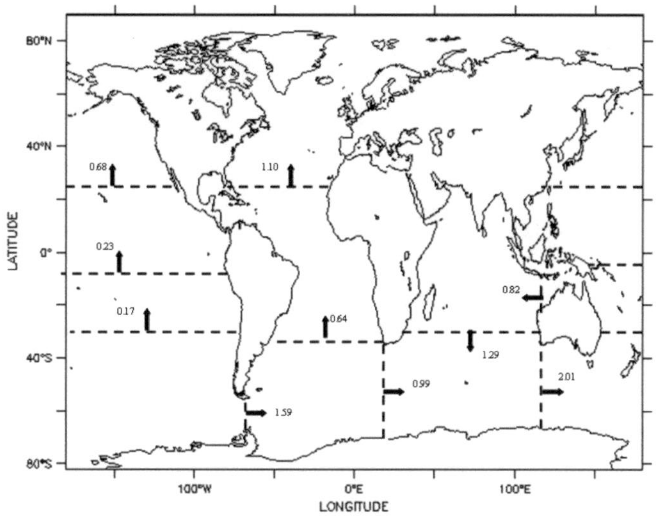

Fig. 12. Annual mean ocean heat transports integrated for the whole water column in each monitored section in the first perturbed integration year.

Figure 13 presents differences between the integrated meridional heat transport of the perturbed (for the first year) and control climate experiments. An intensification of the global meridional overturning cell was observed during this year. An increase of the southward heat transport was observed between 40° S to 10° S, and an increase of the northward heat transport was seen between 7° S to 28° S
The southern oceans show a decrease in the northward heat transport across the South Pacific section, while an increase in the southward heat transport across the 30⁰ South Indian section is observed. These processes could explain the increase of the southward integrated meridional heat transport observed in Figure 14.

Figure 14 shows the whole water column integrated heat transport across each monitored section for the second perturbed integration year. The 32⁰ South Atlantic section exhibited a small decrease in the northward meridional heat transport, while the 25⁰ North Atlantic section exhibited a strong increase of this transport with a value of 1.18 PW. This result seems to be related to the increase in northward volume transport in the upper layer during this year. Also in this year, all the ACC meridional sections presented a decrease of the eastward heat transport associated with the observed decrease in eastward volume transport, with the exception of the Drake Passage deep layer transport, which showed an increase relative to that of the control experiment. As with the first perturbed year, the Indonesian Through flow exhibited a decrease again in eastward heat transport. The southward heat transport across the 30⁰ South Indian section was lower than that observed in the first perturbed integration year, but this value was still greater than observed for the control experiment. The 30⁰ South Pacific and 25⁰ North Pacific sections exhibited an increase in the northward heat transport in comparison with the first perturbed year value.

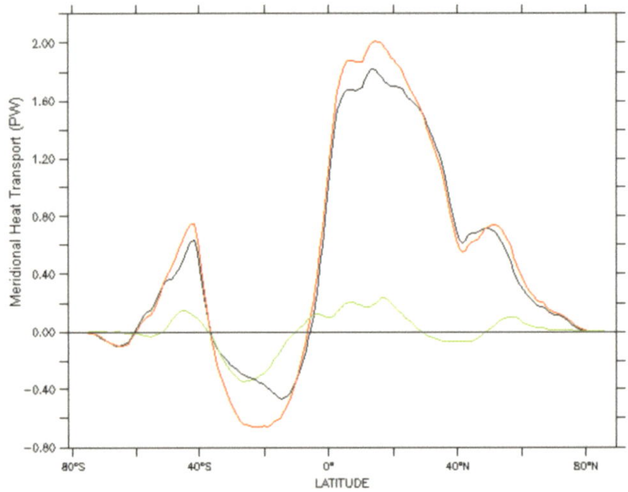

Fig. 13. Global integrated annual mean meridional heat transport (in PW) for the climate experiment (black) and the first year perturbed experiment (red). The difference between the perturbed and climate experiments is also shown (green).

The 10^0 South Pacific section exhibited an inversion of the resultant integrated heat transport from the north to south direction. A value of 0.22 PW was obtained in this section. It is important to note that the second integration year is also the period of greater global wind energy, especially in the Pacific low latitude regions. This fact could explain the inversion of heat transport, which, for this year, also exhibited a strong variance of 1.85 PW^2, which is even greater than the average value.

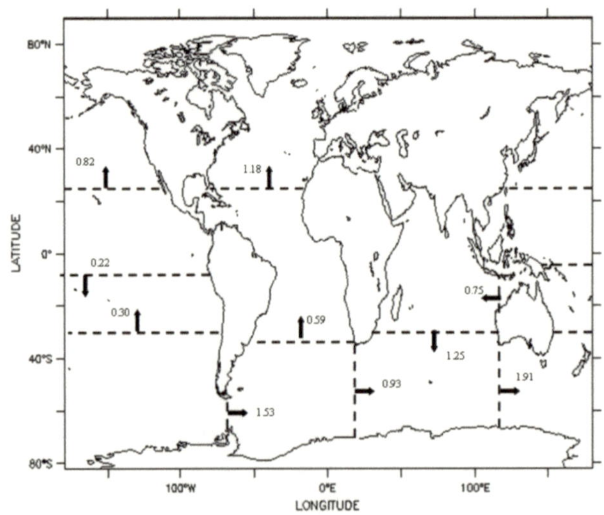

Fig. 14. Annual mean ocean heat transports integrated for the whole water column in each monitored section in the second perturbed integration year.

Figure 15 presents the global integrated meridional heat transport for the second integration year, the control experiment curve and the difference curve between the second perturbed year and the control experiment. This figure reveals a significant increase in the heat transport conveyor belt circulation. An increase of approximately 0.6 PW is observed for the northward heat transport at approximately 20^0 N. An increase of approximately 0.4 PW is also observed in the southward heat transport observed between the equator and 20^0 S. The intensification of this southward heat transport is directly related to the increase in the 30^0 South Indian heat transport and the reversion to the south of the 10^0 South Pacific heat transport.

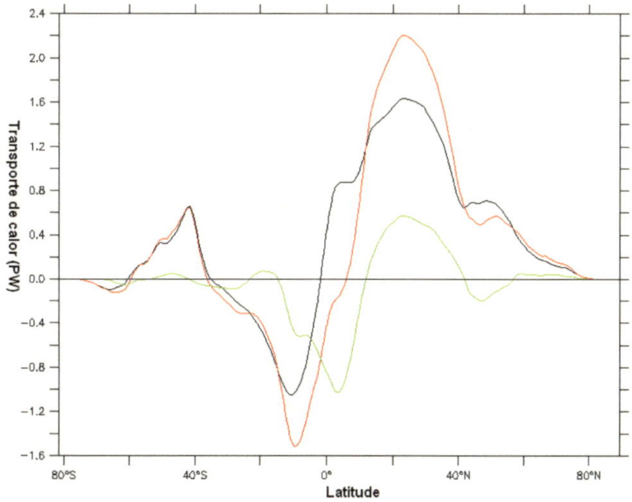

Fig. 15. Global integrated annual mean meridional heat transport (in PW) for the climate experiment (black) and the second year perturbed experiment (red). The difference between the perturbed and climate experiments is also shown (green).

In the third perturbed integration year, the global wind stress field returned slowly to the conditions observed in the first perturbed year. The heat transports obtained in some sections were still greater than the values observed in the control experiment. The whole water column integrated northward heat transport observed in the 25^0 North Atlantic and 32^0 South Atlantic sections were 1.07 PW and 0.66 PW, respectively, for the third year. The Africa-Antarctica, Australia-Antarctica and Drake Passage sections, presented greater eastward heat transport values than the control experiment values. The 30^0 South Indian southward heat transport remains smaller than the value obtained for the control experiment. The Indonesian Throughflow section presented the smallest westward heat transport value obtained for the whole perturbed experiment with a value of 0.65 PW. This result could be associated with the decrease in the upper layer volume transport across this section during this year. The 30^0 South Pacific section exhibited an increase in northward heat transport. The 10^0 South Pacific heat transport reverted to the north, as with its control experiment condition, and it strongly increased in intensity to a value of 0.84 PW. The 25^0 North Pacific section exhibited a decrease in the northward heat transport related to the second integration year, but it was still greater than observed for the control experiment.

The higher meridional integrated heat transport values were observed during the third integration year in the latitudinal band between the equator and approximately 10^0 N. In this region, a maximum value of 2.6 PW for a northward heat transport was observed. This result could be explained by the strong intensification of the northward heat transport at 30^0

and 10^0 in the South Pacific. Northward heat transport also intensified across the 32^0 South Atlantic section, which would contribute to the intensification of the northward integrated heat transport. In the high northern latitudes, there was no apparent difference in heat transport between the perturbed and control experiments.

4. Discussion

The main goal of the control experiment was to represent the important climatological, thermodynamic and dynamic features of global ocean circulation. As demonstrated in section 3, the integrated global ocean kinetic energy shows a well-defined cyclic behaviour starting at the second integration year (Figure 3). The ocean volume and heat transports for the seventh integration year over the monitored sections were also consistent with previous estimates in nearby regions. Even though the analysis of some prognostic model variables, such as the ocean transport field and even the integrated ocean kinetic energy, indicate a cyclic climatic behaviour, it is possible to identify a small time-growth tendency of the kinetic energy curve. This trend is linearly estimated as 1.82×10^{11} Watts/day, which represents only 0.0003 % of the difference between the maximum and minimum global values of the ocean kinetic energy. This result could be attributed to the difference between the OMIP (Ocean Modelling Intercomparison Project) wind stress field used in this work and the ODASI (Ocean Data Assimilation for Seasonal-to-Interannual Prediction) wind stress forcing field that was used to produce the initial conditions (temperature, salinity and ocean velocity fields). According to Wunsch (1998), the largest contribution to the global wind stress field comes from zonal wind stress component, which represents most part of the global work performed by the wind. The main regions of zonal wind energy input are the Southern Ocean, the Kuroshio region, and the Gulf Stream/North Atlantic Current regions.

Figure 16 shows that the zonal wind component in OMIP is more intense than in ODASI at most latitudes, especially for the high latitude bands of both hemispheres, where most of the energetic wind stress field is found. The difference between OMIP and ODASI wind stress globally averaged in space and time is 6.22×10^{11} Watts/day , which is of the same order as the global kinetic energy growth rate (1.82×10^{11} Watts /day).

An important point to be discussed is the methodology used to generate the anomalous ENSO-type global wind stress field used in the perturbed experiment. As cited previously, the perturbed wind stress field was generated by an AGCM (Section 2.2). The boundary condition used for this AGCM was a synthetic sea surface temperature (SST) field that was produced based on the 1982-1983 ENSO event (Torres Jr., 2005). The synthetic SST anomaly field was generated by the combination of two Gaussian analytical functions representing the time and space domains (Figure 17).

Figure 20 reveals that the Gaussian function weight peak is centred in the 16th month, when the most intense SST anomaly values occurred, which are centred between the equator and 15^0 S and between 165^0 W and 90^0 W. An Equatorial Pacific anomaly SST field produced a global wind stress energy peak only in the 21st month and a global ocean kinetic energy peak only in 23rd integration month (Figure 8). There was a delay of seven months between the Equatorial Pacific SST peak and the global ocean kinetic energy peak.

As was observed in the volume and heat transports obtained for the perturbed experiment, changes in the global wind stress field caused by the occurrence of an important global climate variability process such as ENSO could affect the whole world ocean circulation.

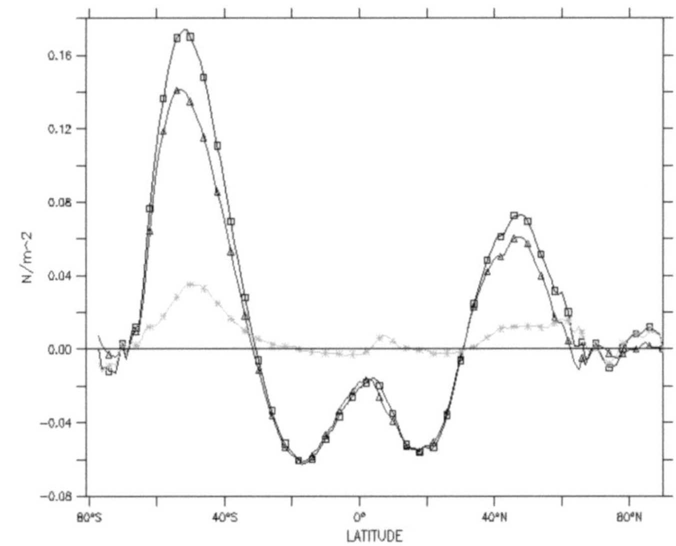

Fig. 16. Zonal distribution of the annually averaged zonal wind stress field from the OMIP (□) and ODASI (Δ) datasets and the difference between them (*).

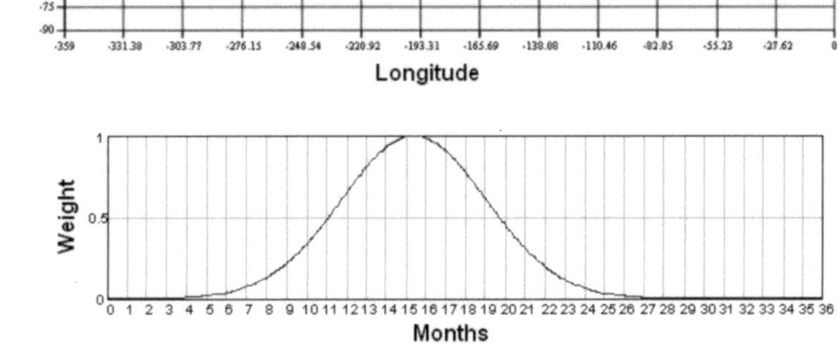

Fig. 17. Space (above) and time (below) Gaussian distributions used to generate the synthetic SST anomaly fields used as the boundary condition of the AGCM.

Figure 18 shows the time series plot of the total heat content in the South Atlantic, North Atlantic, South Pacific and North Pacific basins during the control and perturbed experiments. The black curve represents the cyclic values obtained for the climate experiment for each of these ocean basins. This figure shows for the control experiment a cyclic behaviour with an increase of heat amount during the summer months and a decrease during the winter months as a direct result of the solar radiation cycle, which ensures a gain of heat by the ocean during the summer months and a loss of heat during the winter months. The perturbed experiment revealed for the monitored ocean basins a tendency for heat loss in the northern ocean basins and a tendency for heat gain in the southern ocean basins during the three years of the simulation. These tendencies continued even after the ocean kinetic energy peak that occurred in the 23rd integration month.

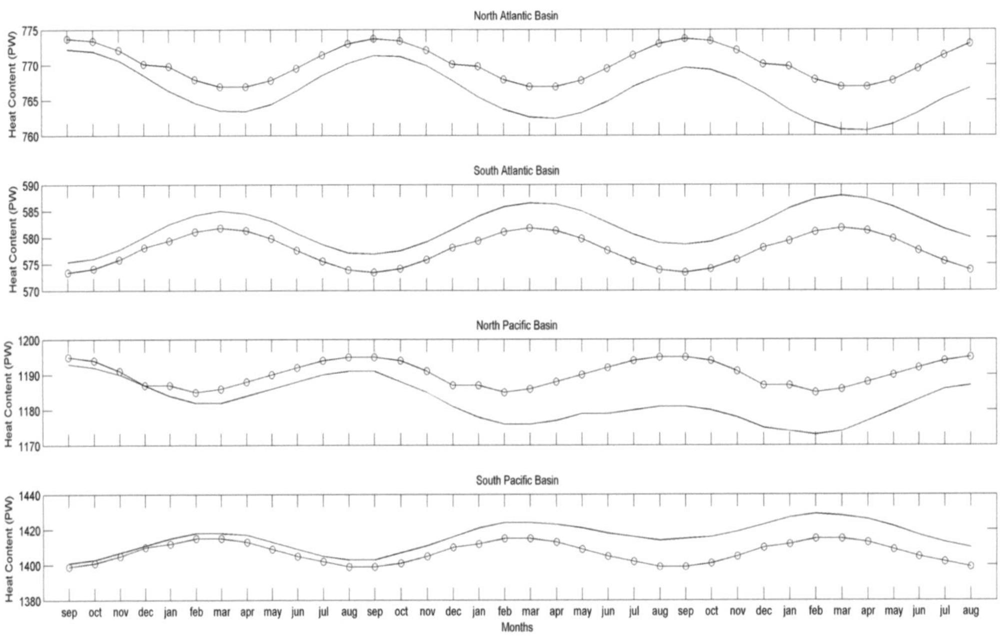

Fig. 18. Time series plot of the total heat content (PW) in the South Atlantic, North Atlantic, North Pacific and South Pacific basins during the perturbed experiment for the control (o) and perturbed experiments.

It is important to note that the only change in boundary conditions for the model in the perturbed experiment was the wind stress field. As described previously, this field was used as an ENSO-type global wind stress field imposed during the three years of the model integration. The sea surface radiation balance is also important to explain the space and time distribution of heat over the ocean basins, but changes in its field typical for ENSO events

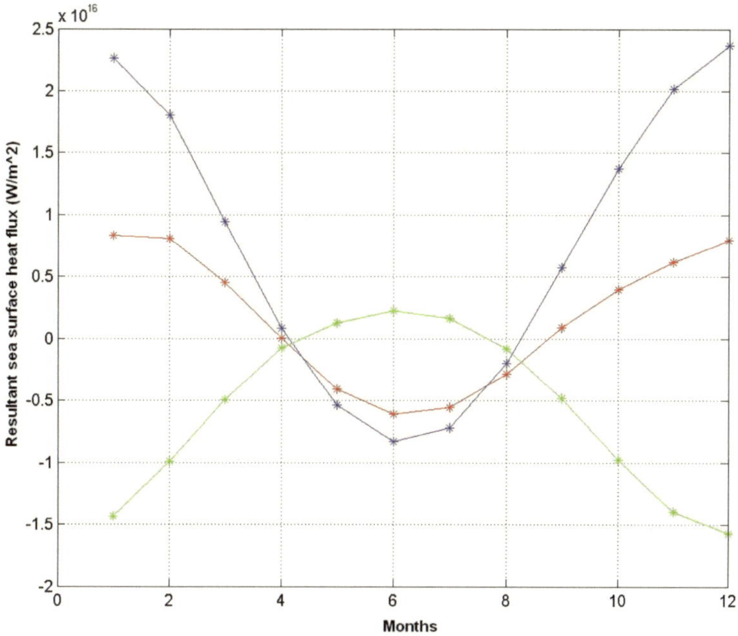

Fig. 19. Control experiment monthly mean resultant sea surface heat flux (W/m2) time series plot. Blue line represents southern hemisphere ocean resultant surface heat flux; green line represents the same property for the northern hemisphere while the red line represents the global resultant surface heat flux.

were not considered in the perturbed experiment. This experiment maintained the climatological radiation fields used in the control experiment. Figure 19 shows the control experiment resultant sea surface heat fluxes for the northern hemisphere, the southern hemisphere and the world ocean areas. This figure reveals the dominance of the southern ocean as a great heat reservoir of the climate system, whereas the northern ocean tends to lose heat to other global reservoirs, such as the atmosphere.

Evidently the sea surface radiation balance varies in function of the ocean atmosphere interaction processes. Not consider this variability constitutes a limitation in the perturbed experiment. For the studied period (three years) these variations would represent little impact over heat and volume ocean transports but this point would be only possible to be verified by the use of an ocean-atmosphere coupled models. These models permit to quantify the amount of heat energy that is exchanged between these two climate components.

Other important point to be mentioned is related with the low meridional and zonal resolution of the model integration grid. This fact avoids the model to represent ocean transports, especially those related with near coast ocean features that must have significant influence in heat and volume transport estimates.

5. Conclusions

This work emphasised the influence of the wind stress field as an important source of ocean climate variability. Although the perturbed experiment used only a three-year duration boundary field, important changes in the global ocean kinetic energy, volume and heat transports due to an ENSO-type global wind stress field were observed. An interval of five months was observed between the SST anomaly peak and the global wind stress energy peak, and an interval of two months was observed between the latter peak and kinetic energy one.

Changes in ocean volume transport were observed not only in the upper ocean layers, which are directly influenced by wind, but also in the deep and bottom density layers. The most significant changes were observed during the second integration year of the perturbed experiment, when the global wind stress energy and, consequently, the ocean kinetic energy exhibited their highest values, respectively 145 PW and 269 PW. A decrease was observed in the upper and deep eastward volume transports associated with the ACC system during the whole perturbed experiment. This result may be related to the weakening of the eastward low level winds that flow around the Antarctic continent. Significant changes in the upper layer meridional Atlantic and Pacific volume transports were also observed. During the second integration year, it was possible to observe an intensification of the northward upper layer volume transport in the 25^0 North Atlantic and 25^0 North Pacific sections, in relation to the apparent increase of the low level subtropical atmospheric gyre circulation. These volume transport changes led to global ocean heat transport changes that could significantly affect the global meridional overturning cell circulation.

The meridional overturning cell changed during the whole perturbed experiment. In the second perturbed integration year, an increase of approximately 0.4 PW of the northward heat transport at 20^0 N and of the same order to the south at 10^0 S was observed. Even though the second integration year was the most energetic, it was in the third simulated year that the highest meridional heat transport value was found, when a northward transport of 2.4 PW was observed between the equator and 20^0 N. These results indicate the occurrence of interannual thermohaline circulation variability despite the short length of the perturbed experiment.

The observed changes in global ocean volume and heat transports could represent an important source of variability to the Earth's climate system and also could contribute in the knowledge of ocean's role in future climate simulations scenarios.

For future works, the use of an ENSO-type transient sea surface radiance balance forcing is recommended with the application of a coupled climate general circulation model with a higher space resolution ocean model. Another important contribution would be the use of wind stress fields from other ENSO events for replicate perturbation experiments.

6. References

Assad, L.P.F.; Torres Jr., A.R.; Candella, R.N.; Mascarenhas Junior, A. S. Brazil Malvinas Confluence Upper Ocean Temperature Anomalies induced by an ENSO wind forcing. Ciencias Marinas, v. 36, p. 267-284, 2010.

Assad LPF, Torres Jr. AR, Zumpichiatti WA, Mascarenhas Jr. AS, Landau L. 2009. Volume and heat Transports in The World Oceans from an Ocean General Circulation Model. Braz. J. Geophys. 27(2): 181-194.

Colberg F, Reason CJC, Rodgers K. 2004: South Atlantic response to El Niño–Southern Oscillation induced climate variability in an OGCM. J. Geophys. Res. 109: C12015, doi:10.1029/2004JC002301.

Ganachaud, A. & C. Wunsch.-2000- Improved estimates of global circulation, heat transport and mixing from hydrographic data. Nature vol: 408: 453 - 457.

Hastenrath, S., 1979. Heat Budget of Tropical Ocean an Atmosphere. Journal of Physical oceanography, 10:159-170.

Lenn, Y-D, Chereskin TK, Sprintall J, Firing E. 2007. Mean jets, mesoscale variability and eddy momentum fluxes in the surface layer of the Antarctic Circumpolar Current in Drake Passage. J. Mar.Res. 65: 27-58.

Lentini CAD, Podestá GG, Campos EJD, Olson DB. 2001. Sea surface temperature anomalies on the western South Atlantic from 1982 to 1994. Cont. Shelf Res. 21:89-112.

Murray, R.J.. 1996. Explicit generation of orthogonal grids for ocean models. Journal of Computational Physics, 126: 251–273.

Oort, A. H., L. A. Anderson AND J. P. PEIXOTO -1994 - Estimates of the energy cycle of the oceans. Journal of Geophysical Research, 99, 7665-7688.

Pacanowsky, R.C. & S.M.Griffies.1999. The MOM3 Manual. Geophysical Fluid Dynamics laboratory/NOAA, Princeton, USA, p. 680.

Rintoul, S.R. 1991. South Atlantic Interbasin Exchange. Journal of Geophysical research, 96, 2675-2692.

Röeske, F. 2001. An Atlas of Surface Fluxes based on the ECMWF Re-Analysis - a Climatological Dataset to force Global Ocean General Circulation Models. Max - Planck Institut für Meteorologie, Hamburg. Report no. 323. ISSN 0937-1060.

Russel,J.L. The Southern Hemisphere Westerlies in a Warming World: Propping Open the Door to the Deep Ocean Journal of Climate. Vol.19. 2006.

Schmitz, W.J.-1996- On The World Ocean Circulation: Volume II. Technical Report. Woods Hole Oceanographic Institution WHOI-96-08.

Stammer, D., C. Wunsch, R.Giering, C. Eckert, P. Heimbach, J.Marotzke, A. Adcroft, C.N.Hill e J.Marshall, 2003. Volume, heat, and freshwater transports of the global ocean circulation 1993-2000, etsimated from a general circulation model constrained by World Ocean circulation Experiment (WOCE) data. Journal of Geophysical Research vol. 108, pp- 7-1 - 7-23.

Sun, C., M. M. Rieneker, A. Rosati, M. Harrison,; A. Wittenberg, C. L. KeppenneEPPENNE, J. P. Jacob, AND R. M. Kovach. 2007. Comparison and sensitivity of ODASI ocean Analyses in the tropical Pacific. Monthly Weather Review, 135: 2242 - 2264.

Torres Jr AR. 2005. Evidências de tele-conexão atmosférica entre fenômenos oceânicos do Pacífico Equatorial e do Atlântico Sul. Tese de Doutorado, Universidade Federal do Rio de Janeiro, Rio de Janeiro, 156 pp.

Wunsch, W. 1998. The Work Done by the Wind on the Oceanic general Circulation. Journal of Physical Oceanography, vol 28: 2332-2340.

Whitworth and Peterson, 1985. T. Whitworth, III and R.G. Peterson, Volume of transport of the Antarctic Circumpolar Current from bottom pressure measurements. Journal of Physical Oceanography 15 (1985), pp. 810–816

Part 4

Rainfall and Drought Assessment

6

Drought Assessment in a Changing Climate

Isabella Bordi and Alfonso Sutera
Department of Physics, Sapienza University of Rome
Italy

1. Introduction

Drought is a natural and recurrent feature of climate. It occurs in all climatic zones, even if its characteristics vary significantly from one region to another, and differs from aridity that is a permanent feature of climate restricted to low rainfall areas. Drought originates from a deficiency of precipitation (less than normal) over an extended period of time, usually a season or more (Wilhite & Glantz, 1985; Bordi & Sutera, 2007). However, since precipitation is related to the amount of water vapor in the atmosphere, combined with the upward forcing of the air masses containing that water vapor, a reduction of either of these may favour the onset of drought conditions. Thus, the phenomenon can be triggered by an above average prevalence of high-pressure systems, by winds carrying continental rather than oceanic air masses, or by high temperatures that enhance evaporation. Moreover, drought is a creeping phenomenon that slowly sneaks up and impacts many sectors of the economy, the environment, and operates on many different time scales (Wilhite et al., 2007). For example, soil moisture conditions respond to precipitation deficits occurring on a relatively short time scale, whereas groundwater, streamflow, and reservoir storage respond to precipitation deficits arising over many months. As a result, drought cannot be viewed solely as a physical phenomenon but it should be considered in relation to its impacts on society. The American Meteorological Society (1997) grouped drought definitions and types into four categories: meteorological, agricultural, hydrological and socioeconomic (Heim, 2002).

Based on all these definitions that nowadays are commonly accepted by the scientific community, three main issues emerge to be important for a comprehensive drought assessment: 1) Development of a drought index able to objectively assess drought conditions of regions characterized by different hydrological regimes and to evaluate the different kinds of droughts, 2) Perform a drought risk analysis in order to prevent the negative impacts of droughts, and 3) Understanding the link between climate variability and drought occurrence in relation to a changing climate. The first point has been addressed first by McKee et al. (1993) by developing the Standardized Precipitation Index (SPI), which is a drought index based only on monthly precipitation. The idea was to quantify the precipitation deficit for multiple time scales that reflect the impacts of drought on the availability of the different water resources. Since the index is standardized through an equal-probability transformation (see next section and Bordi & Sutera, 2004 for details), dry and wet conditions can be monitored in the same way as well as a comparison between locations with different climates is possible. For these reasons the SPI is widely applied for drought monitoring purposes, and in 2009 it has been recommended for characterizing meteorological drought around the world (see Lincon declaration on drought indices, Hayes et al., 2011). Recently, two new standardized drought indices have been proposed for drought variability analysis on multiple time scales, the Reconnaissance Drought index (RDI, Tsakiris et al., 2007) and the Standardized Precipitation Evapotranspiration Index (SPEI, Vicente-Serrano et al., 2010). The indices are based on the supply-demand concept and take into account precipitation (P) and potential evapotranspiration (PET, hence temperature through the Thornthwaite equation). Although both indices comply with the requirement of the standardization, some questions concern their effective capability to better capture temperature changes, and, more importantly, the more appropriate basic variable to be used for drought assessment (i.e. P, P/PET or P–PET, see Raziei et al., 2011).

To address the second point, a full understanding of the third point on climate variability is required. The traditional approach to drought management is reactive, relying largely on crisis management. This approach is usually ineffective because the response comes too late, is poorly coordinated, and costly. In addition, the post-impact response to drought tends to reinforce the existing water resource management methods that often contribute to increase the societal vulnerability to drought. For these reasons, in recent years, governments and institutions involved in water resources management showed more interest in learning how to employ proper risk management techniques to reduce vulnerability to drought and, therefore, lessen the impacts associated with future drought events (Wilhite et al., 2000). In coping with drought following a proactive approach, the first step is the monitoring of the phenomenon and the understanding of the temporal variability of drought events, also in relation to a changing climate (Hayes et al., 2004).

All drought indices developed so far (see the reviews by Heim, 2002 and Keyantash & Dracup, 2002), including the standardized ones mentioned before, are based on drought as a relative concept: drought is defined as the negative departure of meteorological/water-related variables from some pre-established mean conditions (often referred to as calibration period that according to WMO recommendation should be at least 30 years long). By definition, drought occurrence should not depend on slow variations of climate conditions, since a normal climate is defined as the averaged conditions over the long-term record. Unfortunately, data have finite time span so that the stability of the average may undermine an objective assessment, especially when a drift on this random behaviour occurs. Of course, if we were certain that such a drift were of a deterministic kind a few differentiations of the original time series would be in order and would be enough to restore the original definition of climate. However, the situation may be disrupted if the assumptions on the nature of the trend are fallacious or even simply overstated. Moreover, other moments of the climate variables may be very well not stationary. In these cases the entire probability distributions would be affected, and in particular their tails that are, after all, what we are more interested. Things may be even more worsen if variables, relevant for drought occurrence, change differently both in space and time. If, for example, precipitation and temperature in a given location respond differently to climate changes we could overweight one variable with respect to the other reaching unwise conclusions on future drought occurrences.

In the present paper we address the question of how drought indices should be used in a changing climate conditions, i.e. the variables of interest are not stationary. In particular, we evaluate the impact of the observed climate drift on drought assessment. We focus our attention on the European sector providing an updated drought analysis based on the SPI on 24-month time scale computed using the National Centers for Environmental Prediction/National Center for Atmospheric Research (NCEP/NCAR) reanalysis precipitation data from January 1948 to June 2011. The choice of the SPI is motivated by the simplicity and advantages mentioned above, while the long time scale is selected just to filter out high frequency fluctuations in drought signals, highlighting the long-term behaviours to which we are interested on. Moreover, through sample analyses at selected grid points, it is evaluated the existence of a reference calibration period over which the distribution parameters underlying the SPI computation remain stable when a data update is taken into account. In particular, the effect of using a reference calibration period, shorter than the full data record available, on the SPI computation is studied for different climate tendencies. Some conclusions and discussions on future outlooks are provided in the final section.

2. Data and methods

Data used for the analysis are monthly mean precipitation rates retrieved from the NCEP/NCAR reanalysis archive for the period January 1948–June 2011. They are available on the regular grid 1.9°x1.9° in longitude and latitude. Such precipitation data have been derived from the primary meteorological fields of the NCEP medium range forecasting spectral model with 28 "sigma" vertical levels and a triangular truncation of 62 waves, equivalent to about 210-km horizontal resolution. The model is based on the assimilation of a set of observations, such as land surface, ship, rawinsonde, aircraft and satellite data (Kalnay et al., 1996). These data were quality controlled and assimilated with a data assimilation system kept unchanged over the reanalysis period. Though precipitation is not directly assimilated, but derived completely from the model 6-hour forecast, its midlatitude features have been compared favourably with observations and several climatologies (Janowiak et al., 1998; Trenberth & Guillemot, 1998). Since for the present study we have considered the area centred over Europe (25.72°N–71.43°N, 13.13°W–60.00°E), we may feel enough confidence on the data quality.
Hydrological dry/wet conditions over Europe, updated to June 2011, have been assessed through the SPI on 24-month time scale. The SPI computation for a given location and month of the year is based on the long-term precipitation record accumulated over the selected time scale. The empirical probability distribution of the accumulated precipitation is fitted to a theoretical distribution that is then transformed through an equal-probability transformation into a normal distribution. Usually, the two-parameter Gamma distribution is used for fitting the observed precipitation distribution, even if in particular regions other choices may result more suitable (Guttman, 1999). In the present study, we apply the original definition of the SPI by McKee et al. (1993) that is based on the two-parameter Gamma probability density function defined as:

$$g(x) = \frac{1}{\beta^{\alpha}\,\Gamma(\alpha)} x^{\alpha-1} e^{-x/\beta} \quad \text{for } x > 0 \tag{1}$$

where α and β are the shape and scale parameters, respectively, positive defined, x the precipitation amount and

$$\Gamma(\alpha) = \int_0^{\infty} y^{\alpha-1} e^{-y}\, dy \tag{2}$$

with Γ(α) the Gamma function. Even if we recognize that the first source of uncertainty in the SPI computation is the determination of the probability density function, we point out that for our purposes such a choice is not relevant, i.e. the issues addressed here apply regardless of the underlying distribution.
Given the time scale of interest, and for each month of the year, the shape and scale parameters are optimally estimated using the maximum likelihood solutions. The resulting parameters, $\hat{\alpha}$ and $\hat{\beta}$, are then used to find the cumulative probability of an observed precipitation event as:

$$G(x)=\int_0^x g(x)dx=\frac{1}{\hat{\beta}^{\hat{\alpha}}\,\Gamma(\hat{\alpha})}\int_0^x x^{\hat{\alpha}-1}e^{-x/\hat{\beta}}\,dx \qquad (3)$$

Letting $t=x/\hat{\beta}$, equation (3) becomes the incomplete Gamma function

$$G(x)=\frac{1}{\Gamma(\hat{\alpha})}\int_0^x t^{\hat{\alpha}-1}e^{-t}\,dt \qquad (4)$$

Since the Gamma is undefined for x = 0, while a precipitation distribution may contains zeros, the cumulative probability becomes:

$$H(x)=q+(1-q)G(x) \qquad (5)$$

with q the probability of a zero (if n is the number of precipitation observations and m the number of zeros in the precipitation time series, q can be estimated by m/n). The cumulative probability $H(x)$ is then transformed to the standard normal random variable Z with zero mean and unit variance, which is the value of the SPI. By using the approximation provided by Abramowitz and Stegum (1965) that converts cumulative probability to the standard normal random variable Z, we have:

$$\begin{aligned} Z=SPI&=-\left(\tilde{t}-\frac{c_0+c_1\tilde{t}+c_2\tilde{t}^2}{1+d_1\tilde{t}+d_2\tilde{t}^2+d_3\tilde{t}^3}\right) \quad \text{for } 0<H(x)\le 0.5 \\ Z=SPI&=+\left(\tilde{t}-\frac{c_0+c_1\tilde{t}+c_2\tilde{t}^2}{1+d_1\tilde{t}+d_2\tilde{t}^2+d_3\tilde{t}^3}\right) \quad \text{for } 0.5<H(x)<1.0 \end{aligned} \qquad (6)$$

where $\tilde{t}=\sqrt{\ln\left(1/(H(x))^2\right)}$ for $0<H(x)\le 0.5$, $\tilde{t}=\sqrt{\ln\left(1/(1-H(x))^2\right)}$ for $0.5<H(x)<1.0$, while $c_{0,1,2}$ and $d_{1,2,3}$ are constants. Thus, conceptually the SPI represents a z-score, or the number of standard deviations above or below that a precipitation event is from the mean (Bordi & Sutera, 2004). Positive SPI values indicate greater than median precipitation, and negative values indicate less than median precipitation. The SPI classes are defined as: values between -0.99 and +0.99 denote near normal conditions, between –1 and –1.49 moderately dry (hereafter D1), between –1.5 and –1.99 severely dry (hereafter D2) and less than -2 extremely dry (hereafter D3) conditions. The same applies to positive values for wet classes (hereafter W1, W2 and W3 for moderately, severely and extremely wet conditions).

In analyzing the spatial variability of drought across Europe the Principal Component Analysis (PCA) is applied to the SPI field. The PCA consists in computing the covariance matrix of the input data with the corresponding eigenvalues and eigenvectors (Rencher, 1998). The projection of the SPI fields onto the orthonormal eigenfunctions provides the principal components or PC score time series. In guiding a proper interpretation of the results shown in the next section, we remark that the spatial patterns (eigenvectors), properly normalized (divided by their Euclidean norm and multiplied by the square root of the corresponding eigenvalues) are called "loadings" and represent the correlation between the original data (SPI time series at single grid points) and the corresponding PC score time series.

In analyzing the long-term drought variability, as in Bordi et al. (2009) we have evaluated both the linear trends and the leading nonlinear components in the SPI time series. To extract the long-term linear trend we have used the least-squared method to fit a linear model to the time series, while the leading nonlinear components in the SPI time series are extracted using the Singular Spectral Analysis (SSA). SSA technique is a nonparametric spectral estimation method based on embedding a time series in a vector space of dimension M (see Ghil et al., 2002 for details on the technique). In the present study, following Bordi et al. (2009), we have reconstructed the signal considering only the leading component by selecting a window length of $M = 70$ months (i.e. about 1/10th of the time series) because it provides statistically meaningful estimates of the largest resolvable fluctuation period.

3. Results

Firstly, the impact of the actual climate trend (linear and nonlinear) on drought variability in the European sector is assessed. The study is complemented by sample analyses at three selected grid points considered representative of the different climate drifts observed in the available data record. Secondly, the existence of a reference sample size that provides stable estimates of the Gamma distribution parameters, when additional (more recent) data are taken into account, is evaluated.

3.1 The impact of the climate drift on drought variability

We started our analysis by applying the Mann-Kendall test to the NCEP/NCAR monthly precipitation time series from January 1948 to June 2011. The result is shown in Fig. 1a where grey areas denote p-values less than the significance level of 0.01 (i.e. the test rejects the null hypothesis of trend absence). As can be noted, most of northern regions, central Europe and north Africa are characterized by trends of not specified nature, probably affected by seasonality (here not removed). When the accumulated precipitation on 24-month time scale is considered, the statistical test provides the result shown in Fig. 1b. As expected, the area characterized by trend is larger because of the correlation introduced by the accumulation procedure. The application of the statistical test to the SPI on 1-month and 24-month time scales provides the same results (not shown) illustrated in Fig. 1a and 1b, respectively. This means that whenever the precipitation time series has a statistically significant trend, the same holds for the associated SPI on the same time scale.

To investigate which kind of climate drift characterizes the SPI time series in the area of interest, we have applied the PCA to the SPI field. The leading mode of the spatial variability is shown in Fig. 2a, while the associated PC score time series and the fitting linear and nonlinear trends are shown in Fig. 2b. The first loading pattern explains 20.1% of the total variance and seems remained unchanged compared to the one shown by Bordi et al. (2009) in their Fig. 4 where the update was limited to February 2009.

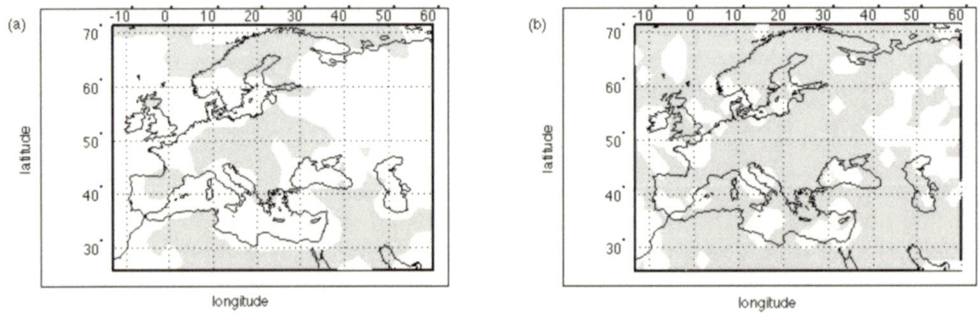

Fig. 1. Results of Mann-Kendall test applied to monthly and accumulated (on 24-month time scale) precipitation time series. Grey areas denote p-values less than the significance level of 0.01.

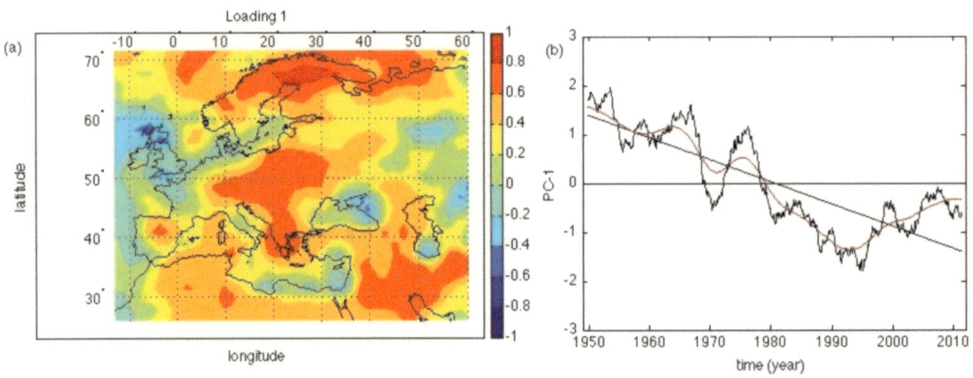

Fig. 2. PCA of the SPI field on 24-month time scale: (a) first loading, (b) First PC score time series and fitting linear and nonlinear trends (black straight line and red line, respectively).

The same holds for the first PC score (PC-1) time series that in the latest two years confirms the fluctuation around -0.5 (i.e. the climate drift occurred from about 1997 onward). The R^2 associated with the linear trend is now 65.4% and the one associated with the reconstructed signal using the leading SSA component is 91.8%. Again, it appears more suitable to represent the long-term SPI variability through a nonlinear fitting instead of the linear one. Thus, the interpretation of the results after this recent update remains basically the same: regions characterized by high positive loading values from the mid seventies onward have experienced prevailing dry events with a change in the last 15 years or so toward near normal conditions (see Bordi et al., 2009 for an in depth discussion). It is worth noting that the PCA technique does not isolate the trend component in the first PC score but just extracts the leading component that maximises the total variance of the SPI field. This means that, for example, the second PC score (not shown) also has a linear trend component, less pronounced, but it is statistically significant at 0.01 level ($R^2 = 5.5\%$).

Now, in analysing the effect of the observed climate drift on drought variability, we have two possibilities: the first one is to remove the observed trend (linear or nonlinear) from the precipitation time series at each grid point, while the second consists in removing the trend on the SPI time series. The first option could provide several problems because of the possible negative values of precipitation derived by subtracting the estimated trend. Thus, it turns out that precipitation (the input variable for the SPI computation) is no more positive defined and the fitting of the empirical precipitation distribution with the two-parameter Gamma is no more possible. Based on the results obtained before (Fig. 1), the second option appears the most suitable. Thus, we continued our analysis by removing the linear and nonlinear trend components from the SPI time series at each grid point, standardizing the SPI residuals time series, and performing again the PCA. Results obtained when the linear trend is removed are shown in Fig. 3a, b, while when the nonlinear trend is excluded are in Fig. 3c, d.

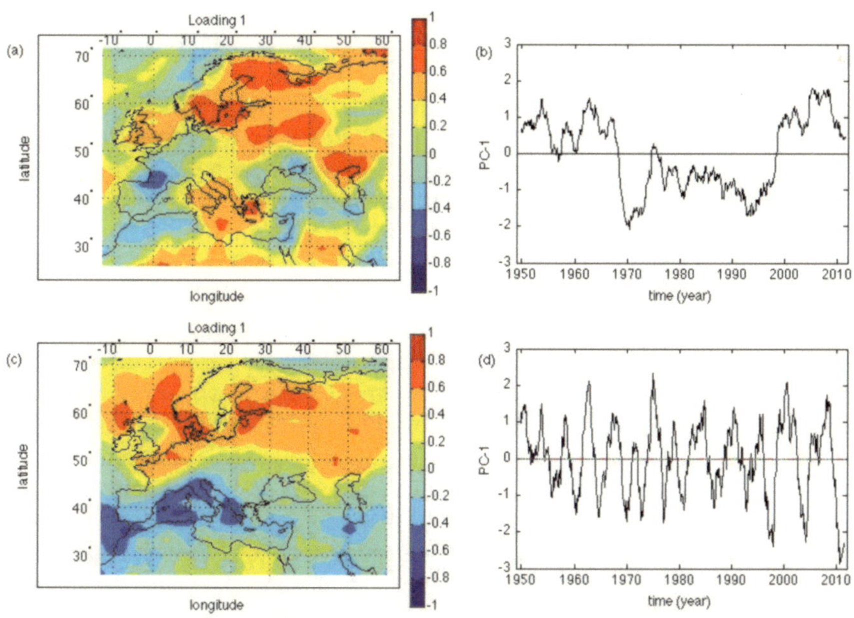

Fig. 3. PCA of the SPI field after removing the linear/nonlinear trend: (a) first loading and (b) first PC score for linear trend removing, (c) first loading and (d) first PC score for nonlinear trend removing.

In case of the removal of linear trend removing the first loading pattern, explaining 14.1% of the total variance, shows high positive values (i.e. high correlations between the standardized residuals of the SPI series and the corresponding PC score) in northern Europe, in the middle of the Mediterranean basin and part of north Africa. The associated PC score has, by construction, no linear trend component, and shows a pronounced minimum around the seventies. Embedded in multi-year fluctuations, three distinct "phases" are noticeable: two periods characterized by positive values, from the beginning of the time record to the seventies and from around 1997 to present, and another one with

negative values in between. It is interesting to compare these results with the ones previously obtained by Bordi and Sutera (2001) for the time section 1950–2000 (see their Panels VI and VII): the addition of the last eleven years has changed remarkably the first loading and the corresponding PC score. This is because the estimated linear trend changed when the time record has been extended. Also, since the reference period used for estimating the Gamma parameters is longer and the precipitation time series are not stationary, the SPI values have been affected by the update (class transition of events).
In case of the removal of nonlinear trend, the first loading explains 13% of the total variance and has a north-south structure. The corresponding PC score is characterized by multi-year fluctuations with a prevalent periodicity around 5.5 years; hence, as expected, the time variability of the SPI residuals is strongly affected by the estimates of the nonlinear trend components (i.e. the choice of M in SSA decomposition). These results suggest that the uncertainty on the nature of trends (linear or nonlinear) leaves undetermined the natural climate variability, which is highly dependent from the assumptions made on the observed tendencies for their estimation.
At this stage of the analysis, it appears interesting to quantify the effect of the estimated climate drift on drought classes. At this purpose, we have computed for selected grid points the percentage of events in each drought/wet class by considering the original SPI time series and the de-trended ones. Due to the high spatial variability of the observed climate trends, we have selected the grid points so that they are representative of no linear trend conditions (North Germany), upward (North England) and downward (Scandinavia) linear trend. Note that the grid points are the same discussed in Bordi et al. (2009) in order to allow the reader to follow the effect of the update of the data subjected to the current climate drift.
Results are illustrated in Fig. 4. In Fig. 4a, b, c we have reported the time behaviour of the precipitation accumulated on 24-month time scale at the three grid points. From a quick look of the bar plots, the different long-term linear trends appear evident, which are the same characterizing the corresponding SPI time series illustrated in Fig. 4d, e, f. In case of linear trend absence (Fig. 4g), we have changes in drought classes only when the first nonlinear SSA component is removed from the SPI time series. The most evident effect is on the severe wet class W2 that changes from 4% to about 7.5%. In case of an upward linear trend (Fig. 4e), accounting for 36% of the total variance of the SPI signal, the impact of removing the trend (both linear and nonlinear) is more evident, especially on moderate and extreme dry/wet classes (Fig. 4h). In the third case (Fig. 4f), the SPI series has high positive correlation with the leading PC score illustrated in Fig. 2b, and the linear downward trend accounts for about 35% of the total variance. The major change (Fig. 4i) occurs for the moderate drought class that increases the percentage of events when the trend is removed; the same holds for the moderate wet class, while for the extreme wet class there is a loss of events when the linear trend is removed. These examples show how the climate drift affects the frequency of occurrence of dry/wet events, showing that it depends on the nature of the estimated trend (linear or nonlinear). In general, it seems that the impact is more pronounced on moderate dry/wet classes (D1 and W1), but taking into account the lower probability of occurrence of the severe and extreme classes, small changes of those events are equally remarkable.

3.2 Reference sample size in a changing climate

Since the computation of the SPI requires the preliminary fitting of a probability distribution to monthly precipitation aggregated at the selected time scale, the index value for a given

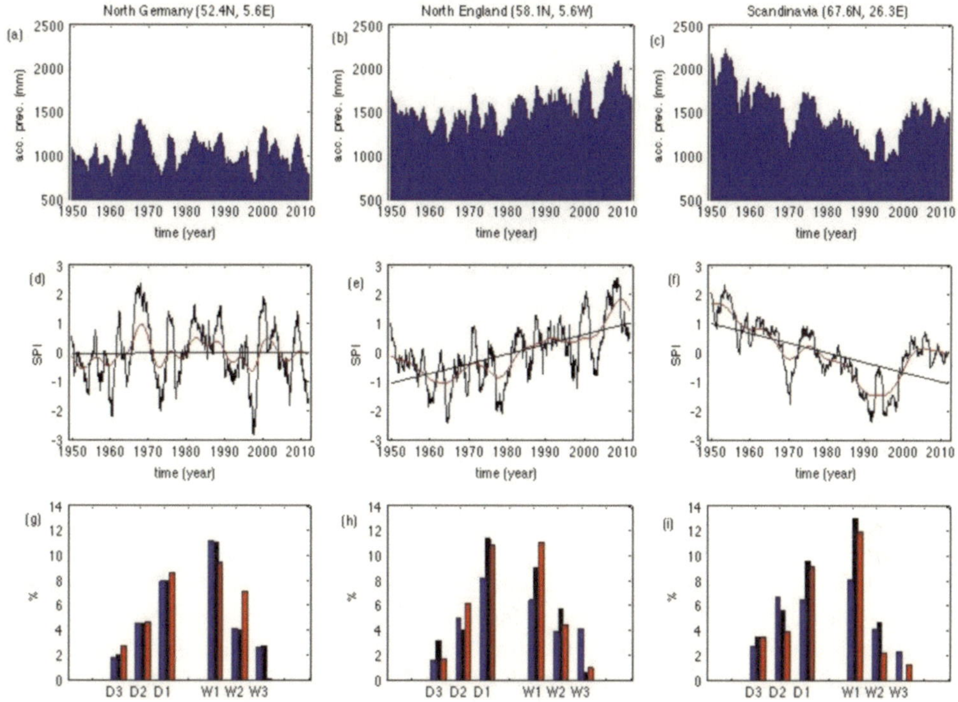

Fig. 4. Accumulated precipitation time series on 24-month time scale at three selected grid points: (a) North Germany, (b) North England, (c) North Europe. SPI time series are in (d), (e) and (f). Straight black lines are the fitting linear trends, red curves the leading SSA components. Bar plots (d)–(f) are the percentages of events in each dry and wet class for the original SPI (blue), the SPI with linear trend removed (black), and the SPI with the leading SSA component removed (red). D1 (W1), D2 (W2) and D3 (W3) stand for moderate, severe and extreme drought (wet) classes, respectively.

year and month will depend on the sample size adopted for its estimation. Usually, the size of precipitation data has to be long enough in order to yield a reliable estimation of the parameters related to the index. McKee et al. (1993) suggested that the length of record for precipitation used in the SPI calculation is "ideally a continuous period of at least 30 years". Guttman (1994) and, later on, Wu et al. (2005) investigated further this problem concluding that longer records are necessary for a stable estimation of the distribution parameters. Unfortunately, the long record requirement of high quality data (observations or reanalysis) cannot be met in most cases, even in the US where precipitation record length varies from one station to another across a region. Moreover, the presence of a trend in the underlying precipitation time series will affect adversely the estimation of the distribution parameters and, therefore, the SPI values. In fact, because the climate is changing, different time periods have different climate mean conditions. In addition, as shown before, the observed trend varies spatially so that the stability of the distribution parameters may be different from one region to another.

In the present subsection we investigate the effect on the SPI computation of the last 15 years, when a change in drought variability is detected. In particular, we address the question whether a reference sample size exists in a changing climate given the actual record length of about 63 years, i.e. the distribution parameters estimated over the reference record remain stable when additional (more recent) data are taken into account in the SPI computation. For the three grid points previously selected as representative of different climate tendencies, we have estimated for each month of the year the parameters (α and β) of the Gamma distributions fitting the accumulated precipitation on 24-month time scale. First, the estimates have been performed using data from 1948–1995 and then 1 year at a time has been added up till 2010 (for a total of 15 years added). Next, the SPI time series for the full record length have been computed using these parameter estimates. This is an application of the "relative" SPI introduced by Dubrovsky et al. (2009). Assuming as the best estimate of α (β) be the one obtained using the full record length, we have computed the percent relative error for each added year as:

$$\varepsilon_\alpha = \frac{\alpha_i - \alpha_N}{\alpha_N} \tag{3}$$

with $i = 1, 2, \ldots N$, and $N = 15$. The same holds for β. Voluntarily, the absolute value has not been applied to ($\alpha_i - \alpha_N$) in order to retain the information on the decreasing/increasing behaviour of the Gamma parameters as a function of time. In Fig. 5, for the three selected grid points, the relative error (in %) on α and β, the SPI time mean and the standard deviation as a function of the number of years added to estimate the distribution parameters are shown.

From the figure it can be noted that for North Germany (absence of statistically significant linear trend in the SPI series), the relative errors on the Gamma parameters are bounded at 10%, and the mean and the standard deviation of the relative SPI are close to 0 and 1, respectively. This result shows that the distribution parameters remain almost stable in the latest 15 years, resulting in a preservation of the basic characteristics of the SPI. Thus, in this case the period 1948–1995 can be taken as a reference calibration period for assessing recent climate conditions; this is because the climate mean conditions had not remarkably changed in the latest 15 years. In case of North England, instead, where an upward linear trend has been detected in the SPI time series, the relative errors on the Gamma parameters are higher and cross the threshold of 10% just when almost all the recent 15 years are taken into account. The mean and standard deviation of the relative SPI strongly depart from 0 and 1, respectively. This suggests that in this case the whole precipitation record must be considered in computing the SPI, and a shorter record is not suitable for assessing current drought conditions. Lastly, for the grid point located in Scandinavia, where the SPI showed a downward linear trend embedded in a long-period fluctuation, results suggest that a suitable calibration period for the computation of the relative SPI may be 1948–2005, i.e. by excluding the latest five years no remarkable errors occur in the Gamma parameters estimates.

In applying the relative index, another problem arises: even small changes in the distribution parameter estimates (relative errors below 10%) might change the number of dry/wet events falling in each SPI dry/wet class. To quantify this effect we have reported in Table 1 the percentage of dry/wet events for the calibration period 1948–2010 (full record) and 1948–1996 (short record). It is worth noting that when a very long time record is considered, by construction, each SPI class corresponds to a known probability (for example, extreme dry events occur with probability 2.3%). Here, the deviations from the nominal probability values are due to the limited record size.

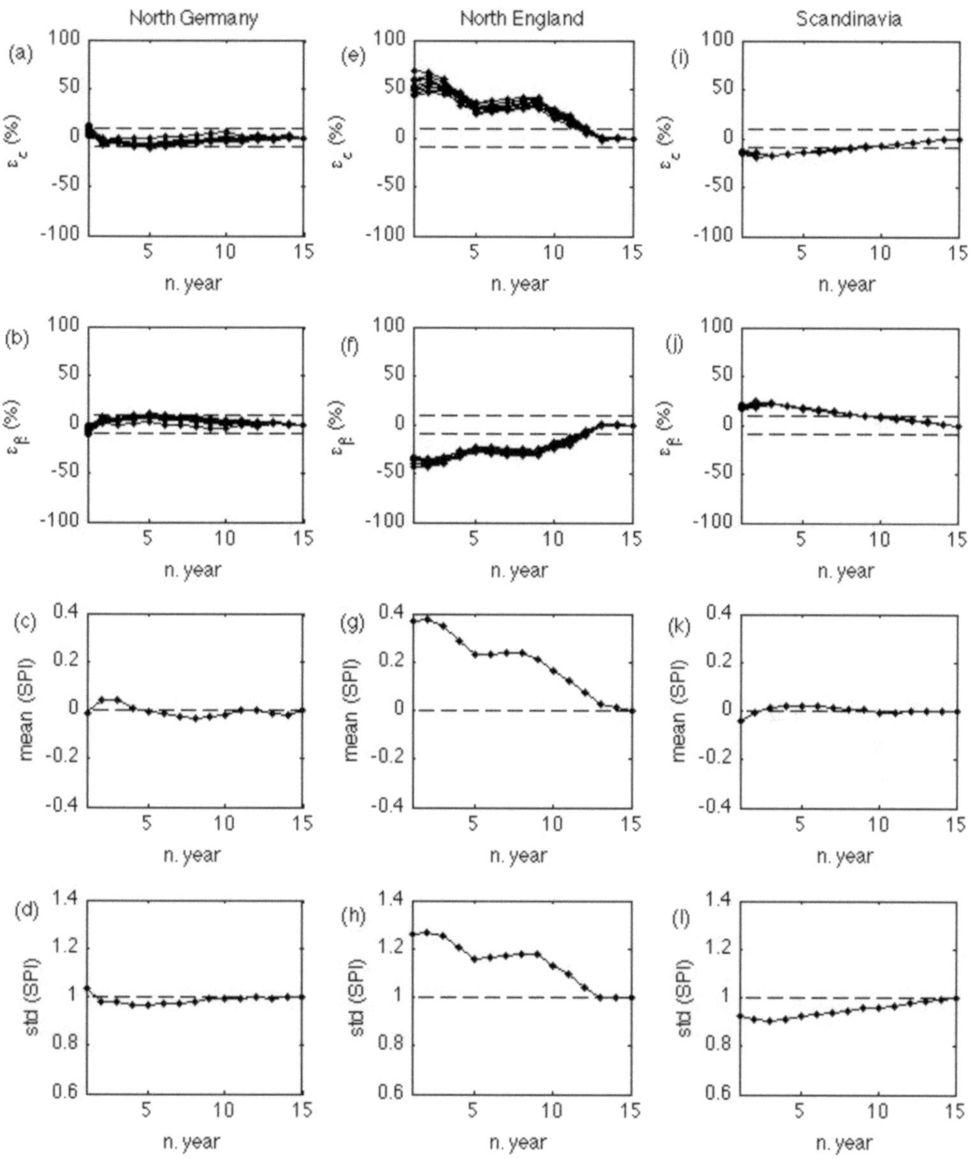

Fig. 5. For the three selected grid points: percent relative error on α and β for each month of the year, SPI time mean and standard deviation as a function of the number of years added for the Gamma parameter estimates. Horizontal dashed lines on ε_α and ε_β time behaviours denote the +10% and –10% levels.

	North Germany		Norh England		Scadinavia	
Dry/wet classes	Full record	Short record	Full record	Short record	Full record	Short record
D3 (extreme dry)	2.2	2.3	1.6	2.3	2.7	1.0
D2 (severe dry)	4.1	4.6	5.0	4.6	7.1	7.0
D1 (moderate dry)	8.5	8.6	8.3	7.1	6.5	8.0
W1 (moderate wet)	11.1	10.5	6.5	11.5	8.2	6.5
W2 (severe wet)	4.2	4.6	4.0	6.7	4.1	5.5
W3 (extreme wet)	2.7	3.0	4.2	10.2	2.5	0.1

Table 1. Percentage of dry/wet events in each class when the SPI at the three reference grid points (North Germany, North England, Scandinavia) is computed using the full record length of data or the shorter period 1948–1996 for estimating the Gamma parameters.

Table 1 shows that for North Germany there is only a slight change of the percentage of events in each class, while for the other two grid points the differences are more evident. To be point out is the great increase of extreme wet events in North England, while the opposite occurs in Scandinavia. The impact on dry classes seems less dramatic.

To better understand these changes we show in Fig. 6 the monthly precipitation aggregated on 24-month time scale as a function of the corresponding SPI values for the two calibration periods: full record (blue) and shorter record (1948–1996, red). The dispersion of the points in the plots is related to the month-to-month variability within the year (SPI values are calibrated separately for each month). It is interesting to note how the precipitation threshold levels associated with dry/wet classes changed noticeably in the three locations. For example, while for the grid point in North Germany there are no appreciable changes, for the grid point in North England there are noticeable differences, especially for the wet classes. The threshold level on the accumulated precipitation for extreme wet conditions (SPI > 2) is about 1950 mm when the calibration period is the full record, while it is reduced to about 1800 mm when the calibration period is shorter (1948–1996). This is the reason why the relative SPI computed on the shorter time period provides a higher percentage of extreme wet events. Similar considerations can be made for the other classes.

Summarizing, since climate conditions vary spatially and temporally, and the climate tendencies (for example, long-term linear trends) highly depend on the record length considered, there is not a general rule for establishing whether a reference sample size, shorter than the full data record available, exists and can be used as a calibration period for the SPI computation. Such a calibration period should be long enough to ensure the stability

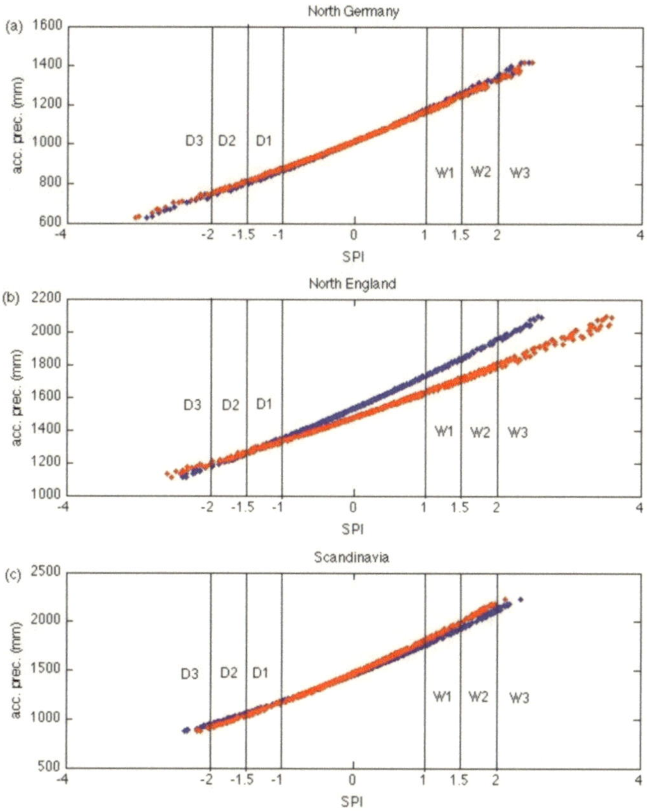

Fig. 6. Accumulated precipitation on 24-month time scale as a function of the corresponding SPI values for the three selected grid points. The SPI Gamma parameters have been estimated using the full precipitation record (blue) and the shorter period 1948-1996 (red). Vertical black lines denote the different dry/wet classes. Units of precipitation are mm.

of the Gamma distribution parameters underlying the index computation even when additional (more recent) data are taken into account. The methodology here proposed, and illustrated through a few examples, may be of reference for addressing the problem in a changing climate conditions. First of all, it is preferable to provide a regionalization of drought conditions in the area of interest, identifying sub-regions characterized by independent drought variability. At this purpose, the application of the PCA with VARIMAX rotation can be a useful tool (see for example Bordi & Sutera, 2001, Raziei et al., 2010 and references therein), or, as suggested by Hannachi et al. (2009), the alternative rotation method based on the Independent Component Analysis (ICA) can be used. Then, for each sub-region a study of the stability of the Gamma parameters can be carried out as a function of the record length. As a general rule, and as illustrated before, it is expected that the SPI time series not characterized by statistically significant long-term linear trends are not much affected by the update of recent data, provided that the reference sample size is long enough to yield a reliable estimation of the parameters. Differently, the SPI at sub-regions characterized by upward or downward long-term linear trends may be quite affected by the update. However, a shorter (with respect to the full record length available) reference period can be found depending on the long-term structure of the SPI tendency. The different results obtained for the grid points at North England and Scandinavia, in fact, suggest that the presence of a linear trend in the SPI series does not exclude the possibility to define a shorter reference calibration period. Moreover, whenever a reference period is defined and the relative SPI applied, a check on the percentage of events falling in each dry/wet class is desirable, to quantify even the effect of small differences between the SPI values computed using the full record and the shorter one. Lastly, the impact of the data update on the precipitation thresholds associated with the SPI dry/wet classes (Fig. 6) appears of interest to quantify the recent changes of climate conditions.

4. Conclusion

In the present paper we have illustrated some approaches for assessing drought conditions in a changing climate. The analysis has been focused on the European sector using the NCEP/NCAR reanalysis precipitation data from 1948 to 2011 (June). This is for continuity with the previous study by Bordi et al. (2009) where the authors analysed the recent changes occurring in drought and wetness conditions over Europe after the update to 2009.
Since for drought analysis a standardized and multi-scale drought index is preferred to facilitate the quantitative comparison of drought events at different locations and time scales, we chose the SPI on 24-month time scale for our study. The long time scale is selected to filter out high frequency fluctuations in drought signals, highlighting long-term behaviours to which we are interested on. Moreover, we applied the original definition of the SPI based on the fitting of the empirical accumulated precipitation distributions with the two-parameter Gamma function. It is worth noting that our arguments are independent on the theoretical probability density function underlying the SPI computation.
After verifying (through the Mann-Kendall test) that if there is a trend in precipitation time series at a given location the same holds for the associated SPI, we have performed the PCA of the SPI field and analysed the leading spatial mode and temporal principal component score. As expected, the first loading remained the same after the recent update (only two more years) and the corresponding PC score confirmed the actual tendency toward near normal conditions in the areas characterized by high positive loading values. Then, the long-term tendency of the SPI time series at each grid point is estimated through a linear and nonlinear fitting, and removed from the original index signals. The resulting SPI residuals were standardized and decomposed into principal components. The first loading and PC score obtained represent the leading mode of space and time drought variability when the estimated climate drift is removed. It turns out that the results are very different, dependent on the nature of the trend (linear or nonlinear). Moreover, in case of the nonlinear trend, here estimated through the SSA with a time window M of 70 months, the oscillatory behaviour of the first PC score depends on the choice of M. For three sample grid points, considered representative of climate conditions with not statistically significant linear trend and statistically significant upward and downward trend, we have quantified the class change of dry/wet events due to the trend removal. The key studies show how the climate drift affects the frequency of occurrence of dry/wet events, showing that it depends on the nature of the estimated trend. It appears that the impact is more pronounced on moderate dry/wet classes but, taking into account the lower probability of severe and extreme events, small changes of those events are equally remarkable.

In the second part of the paper we have addressed the question of the reference sample size useful for the computation of the relative SPI proposed by Dubrovsky et al. (2009). For the three grid points previously considered, we have estimated for each month of the year the parameters α and β of the Gamma distributions fitting the monthly precipitation accumulated on 24-month time scale. The estimates have been performed using data from 1948–1995 and then 1 year at a time has been added up till 2010 (for a total of 15 years added). Using such parameter estimates, the SPI time series for the full record length have been computed. Results suggest that in case of linear trend absence, the period 1948–1995 can be taken as a reference calibration period because the climate mean conditions, and hence α and β, had not remarkably changed in the last 15 years. For the second grid point considered, instead, the whole precipitation record must be taken into account for properly assessing recent climate conditions. Lastly, for the third grid point, a suitable calibration period is estimated to be 1948–2005. Thus, the different results obtained for the latter grid points (North England and Scandinavia) suggest that the presence of a linear trend in the SPI series does not exclude the possibility to define a shorter calibration period.

The impact of the data update on the precipitation thresholds associated with the SPI dry/wet classes (Fig. 6) appears of interest to quantify the recent changes of climate conditions. It is worth noting that this approach, also finds a useful application in case of drought analysis in a different climate, i.e. comparison of dry/wet conditions between the control run of a given general circulation model and the corresponding scenario run. This kind of analysis is nowadays very common due to the increasing interest in future climate projections related to the increasing greenhouse gases emission (see for example Burke & Brown, 2008; Sheffield & Wood, 2008; Bothe et al., 2011). In dealing with this issue by using a standardized drought index, like the SPI, there are three viable strategies: compute the SPI using the full data record (control plus scenario), compute the relative SPI (relative to the control climate mean conditions), or compute the SPI separately for the control and future run. Since, by construction, the future climate is expected to be different from the control, the first choice will change drastically the SPI values computed for the control (present climate), without providing useful information. The second option, based on the relative SPI, has the shortcomings discussed above (i.e. the index does not preserve the zero mean and unit variance) and its interpretation can be misleading since the dry/wet classes refer to the Gamma distributions of the present climate, kept (wrongly) unchanged in the future. In other words, the loss of the standardization of the SPI is due to the erroneous fits of the actual precipitation distributions. The third option, instead, offers the possibility to properly compute the SPI in the two time sections, but now the index classes are relative to two different climate mean conditions. Thus, in this case the mapping of the accumulated precipitation versus the SPI values, both for the control and the projection, can provide an estimation of the new precipitation thresholds associated with dry/wet events. These thresholds may be then used to estimate the return times of extreme events when a point over threshold method is applied (see Bordi et al., 2007 and references therein).

In concluding, based on the high spatial variability of the observed climate trends and its dependence on the time section considered (usually too short for a comprehensive drought assessment), future efforts should be devoted to the analysis of synthetic precipitation time series where the nature of the trend is prescribed, with no limitation on the record length. In this way, the climate drift is known and its impact on drought can be studied applying different drought indices, testing their suitability under various changing climate assumptions (i.e. for example, change of the mean or variance in the precipitation distributions). In particular, it should be interesting to start from the observed climate conditions in the last 60 years or so and simulate possible future evolutions of the variable of interest (in our case the precipitation). Then, perform a risk analysis associated with the underlying assumption on the climate trend. This will be the topic of a next study.

5. Acknowledgment

NCEP/NCAR data have been freely provided by the NOAACIRES Climate Diagnostic Center, Boulder, Colorado, from their web site at http://www.cdc.noaa.gov.

6. References

Abramowitz, M. & Stegum, I. A. (1965). *Handbook of Mathematical Functions*, Dover Pub., ISBN 0-486-61272-4, New York, USA.

American Meteorological Society (AMS) (1997). Meteorological Drought-Policy Statement. *Bulletin of American Meteorological Society*, Vol.78, No.5, (May 1997), pp. 847–849, ISSN 0003-0007.

Bordi, I. & Sutera, A. (2001). Fifty Years of Precipitation: Some Spatially Remote Teleconnections. *Water Resources Management*, Vol.15, No.4, (August 2001), pp. 247–280, ISSN 0920-4741.

Bordi I. & Sutera, A. (2004). Drought Variability and its Climatic Implications. *Global and Planetary Change*, Vol.40, No.1, (January 2004), pp. 115–127, ISSN 0921-8181.

Bordi, I. & Sutera, A. (2007). Drought Monitoring and Forecasting at Large-Scale, In: *Methods and Tools for Drought Analysis and Management*, G. Rossi, T. Vega and B. Bonaccorso, (Eds.), pp. 3–27, Springer-Verlag New York Inc., ISBN 13: 9781402059230, New York, US.

Bordi, I., Fraedrich, K., Petitta, M. & Sutera, A. (2007). Extreme Value Analysis of Wet and Dry periods in Sicily. *Theoretical and Applied Climatology*, Vol.87, No.1-4, (January 2007), pp. 61–71, ISSN 0177-798X.

Bordi, I. ; Fraedrich, K. & Sutera, A. (2009). Observed Drought and Wetness Trends in Europe : An Update. *Hydrology and Earth Systems Sciences*, Vol.13, No.8, (August 2009), pp. 1519-1530, ISSN 1027-5606.

Bothe, O.; Fraedrich, K. & Zhu, X. (2011). Large-Scale Circulations and Tibetan Plateau Summer Drought and Wetness in a High-Resolution Climate Model. *International Journal of Climatology*, Vol.31, No.6, (May 2011), pp. 832–846, ISSN 0899-8418.

Burke, E.J. & Brown, S.J. (2008). Evaluating Uncertainties in the Projection of Future Drought. *Journal of Hydrometeorology*, Vol.9, No.2, (April 2008), pp. 292–299, ISSN 1525-755X.

Dubrovsky, M.; Svoboda, M.D.; Trnka, M.; Hayes, M.J.; Wilhite, D.A.; Zalud, Z. & Hlavinka, P. (2009). Application of Relative Drought Indices in Assessing Clmate-Change Impacts on Drought Conditions in Czechia. *Theoretical and Applied Climatology*, Vol.96, No.1-2, (April 2009), pp. 155–171, ISSN 0177-798X.

Ghil, M.; Allen, M.R.; Dettinger, M.D.; Ide, K.; Kondrashow, D.; Mann, M.E.; Robertson, A.W.; Saunders, A.; Tian, Y.; Varadi, F. & Yiou, P. (2002). Advanced Spectral Methods for Climatic Time Series. *Reviews of Geophysics*, Vol.40, No.1, (September 2002), pp. 3.1–3.41, ISSN 8755-1209.

Guttman, N.B. (1994). On the Sensitivity of Sample L Moments to Sample Size. *Journal of Climate*, Vol.7, No.6, (June 1994), pp. 1026–1029, ISSN 0894-8755.

Guttman, N.B. (1999). Accepting the Standardized Precipitation Index : A Calculation Algorithm. *Journal of the American Water Resources Association*, Vol.35, No.2, (April 1999), pp. 311–322, ISSN 1093-474X.

Hannachi, A.; Unkel, S.; Trendafilov, N.T. & Jolliffe, I.T. (2009). Independent Component Analysis of Climate Data: A New Look at EOF Rotation. *Journal of Climate*, Vol.22, No.11, (June 2009), pp. 2797–2812, ISSN 0894-8755.

Hayes, M.J.; Wilhelmi, O.V. & Knutson, C.L. (2004). Reducing Drought Risk: Bridging Theory and Practice. *Natural Hazards Review*, Vol.5, No.2, (May 2004), pp. 106–113, ISSN 1527-6988.

Hayes, M.; Svoboda, M.; Wall, N. & Widhalm, M. (2011). The Lincon Declaration on Drought Indices: Universal M eteorological Drought Index Recommended. *Bulletin of American Meteorological Society*, Vol.92, No.4, (April 2011), pp. 485–488, ISSN 0003-0007.

Heim Jr, R.R. (2002). A Review of Twentieth-Century Drought Indices used in the United States. *Bulletin of the American Meteorological Society*, Vol.83, No.8, (August 2002), pp. 1149–1165, ISSN 0003-0007.

Janowiak, J.E.; Gruber, A.; Kondragunta, C.R., Livezey, R.E. & Huffman, G.J. (1998). A Comparison of the NCEP-NCAR reanalysis Precipitation and the GPCP Rain Gauge-Satellite Combined Dataset with Observational Error Considerations. *Journal of Climate*, Vol.11, No.11, (November 1998), pp. 2960–2979, ISSN 0894-8755.

Kalnay, E.; Kanamitsu M.; Kistler R.; Collins W.; Deaven, D.; Gandin L.; Iredell M.; Saha, S.; White, G.; Woollen, J.; Zhu, Y.; Leetmaa, A.; Reynolds, R.; Chelliah, M.; Ebisuzaki, W.; Higgins W.; Janowiak, J.; Mo, K.C.; Ropelewski, C.; Wang, J.; Jenne, R. & Joseph, D. (1996). The NCEP/NCAR 40-year Reanalysis project. *Bulletin of American Meteorological Society*, Vol.77, No.3, (March 1996), pp. 437–471, ISSN 0003-0007.

Keyantash, J. & Dracup, J.A. (2002). The Quantification of Drought : An Evaluation of Drought Indices. *Bulletin of American Meteorological Society*, Vol.83, No.8, (August 2002), pp. 1167–1180, ISSN 0003-0007.

McKee, T.B. ; Doesken, N.J. & Kleist, J. (1993). The Relationship of Drought Frequency and Duration to Time Scales, *Proceedings of the AMS 8th Conference on Applied Climatology*, pp. 179–184, Anaheim, CA, January 17–22, 1993.

Raziei, T.; Bordi, I.; Pereira, L.S. & Sutera A. (2010). Space-Time Variability of Hydrological Drought and Wetness in Iran Using NCEP/NCAR and GPCC Datasets. *Hydrology and Earth System Sciences*, Vol.14, No.10, (October 2010), pp. 1919–1930, ISSN 1027-5606.

Raziei, T.; Bordi, I.; Pereira, L.S. & Sutera A. (2011). Detecting Impacts of a Changing Climate on Drought Characteristics in Iran, *Proceedings of the VI International Symposium of EWRA on Water Engineering and Management in a Changing Environment*, Catania, Italy, June 29–July 2, 2011, ISSN 2038-5854.

Rencher, A.C. (1998). *Multivariate Statistical Inference and Applications*, John Wiley & Sons Inc., ISBN 0-471-57151-2, New York, USA.

Sheffield, J. & Wood, E.F. (2008). Projected Changes in Drought Occurrence Under Future Global Warming from Multi-Model, Multi-Scenario, IPCC AR4 Simulations. *Climate Dynamics*, Vol.31, No.1, (July 2008), pp.79–105, ISSN 0930-7575.

Trenbert, K.E. & Guillemot, C.J. (1998). Evaluation of the Atmospheric Moisture and Hydrological Cycle in the NCEP/NACR reanalyses. *Climate Dynamics*, Vol.14, No.3, (March 1998), pp. 213–231.

Tsakiris, G.; Pangalou, D. & Vangelis, H. (2007). Regional Drought Assessment Based on the Reconnaissance Drought Index (RDI). *Water Resources Management*, Vol.21, No.5, (May 2007), 821–833, ISSN 0920-4741.

Vicente-Serrano, S.M.; Buguería, S. & Lòpez-Moreno, J.I. (2010). A multiscalar Drought Index Sensitive to Global Warming: The Standardized Precipitation Evapotranspiration Index. *Journal of Climate*, Vol.23, No.7, (April 2010), pp. 1696–1718, ISSN 0894-8755.

Wilhite, D.A. & Glantz, M.H. (1985). Understanding the Drought Phenomenon: The Role of Definitions. Water International, Vol.10, No.3, (available online 22 January 2009), pp. 111–120, ISSN 0250-8060.

Wilhite, D.A. ; Hayes, M.J. ; Knutson, C. & Smith, K.H. (2000). Planning for Drought : Moving from Crisis to Risk Management. *Journal of the American Water Resources Association*, Vol.36, No.4, (August 2000), pp. 697–710, ISSN 1093-474X.

Wilhite, D.A. ; Svoboda, M.D. & Hayes, M.J. (2007). Understanding the Complex Impacts of Drought: A Key to Enhancing Drought Mitigation and Preparedness. *Water Resources Management*, Vol.21, No.5, (May 2007), pp. 763–774, ISSN 0920-4741.

Wu, H. ; Hayes, M.J. ; Wilhite, D.A. & Svoboda, M.D. (2005). The Effect of the Length of Record on the Standardized Precipitation Index Calculation. *International Journal of Climatology*, Vol.25, No.4, (March 2005), pp. 505–520, ISSN 0899-8418.

7

Seasonal Summer Rainfall Prediction in Bermejo River Basin in Argentina

Marcela H. González and Ana María Murgida
Universidad de Buenos Aires
Argentina

1. Introduction

Bermejo River Basin is located in the Chaco Plains in northern Argentina (Fig 1). The river has an extension of 1,450 km and the basin area covers 16,048 km^2, comprising the north of Salta and the Formosa and Chaco provinces. Its principal tributary is San Francisco River which carries mountain waters. Two different sections can be detected in Bermejo River: the upper and the middle-low Bermejo. Vegetation is wooded with more plains to the east and with the presence of isolated yungas. Vast areas have been historically inhabited by indigenous communities with extensive farming practice. Not taking into account further agricultural colonization, traditional productive activities in the dry Chaco were based on the supply of forest resources and rivers: timber and firewood, cattle grazing and goats, hunting and commercial and subsistence fishing, and harvesting of fruits (carob, mistol), fibers (chaguar) and honey. Nowadays, economic activity is typically agricultural and it is located in two main centres separated by a large plain: east of Salta and east of Chaco. Moreover, historical data show that the region has been the scene of frequent hydro-meteorological disasters (floods and droughts) and the impacts of these events have had a strong impact on the welfare of the population, productive activities and infrastructure. Fifty years ago, agricultural enterprises were risky. Today they are a tool for development and a forcing factor for vulnerabilization of productive instruments and traditional living. There is ample evidence that climate change impacts are already being observed today and that policies that seek the best ways to meet them are essential for the development and welfare of the community. Therefore, the Argentinean Chaco is a region that, as a result of the change in land use, presents "hotspot "or critical areas for the period 1982-1999 (Baldi et al 2008; http://lechusa.unsl.edu.ar /). They are the result of the implementation of deforestation for the advancement of agriculture and intensive farming.

In this region the climate is subtropical with a mean annual rainfall cycle showing a minimum in winter, which is more pronounced in the west, with dry conditions prevailing from May to September (González and Barros 1998, Reboita et al 2010). The Andes chain lies along the west of Argentina and prevents the access of humidity from the Pacific Ocean. Therefore, the flow is governed by the South Atlantic High and as a consequence, winds prevail from the north and the east. An intermittent low pressure system, whose origin could be a combination of thermal and dynamical effects, is located between 20° and 30°S, in a dry and relatively high area east of the Andes. This system is observed all year long, though it is deeper in summer than in winter. When this low is present, northerly flow is favoured at low levels over the subtropical region. Therefore, the water vapour entering at low levels comes from both the tropical continent and the Atlantic Ocean. In the first case, the easterly low-level flow at low latitudes is channelled towards the south between the Bolivian Plateau and the Brazilian Planalto, advecting warm and humid air to southern Brazil, Paraguay, Uruguay and subtropical Argentina and depicting a typical feature that many authors have studied (Lenters and Cook 1995, Wang and Paegle 1996, Barros et al 2002 and Vera et al 2006). Often, an intense low-level jet develops and it enhances humid and warm air advection, especially in summer (Douglas et al. 1998, Paegle 2000, Silva Dias 2000, Salio et al 2002). Intermittent eruptions of polar fronts from the south modify this picture, causing a west or a southwest flow in low levels after the frontal passage.

The great interanual rainfall variability generates the requirement to understand the large circulation patterns associated with different hydric situations. Some remote sources affect the before-mentioned interannual variability. Subtropical South America is known to be one of the world regions with an important El Niño-Southern Oscillation (ENSO) signal in the precipitation field (Ropelewski and Halpert 1987, Kiladis and Diaz 1989). This signal varies along each of the ENSO phases, and it differs among regions (Grimm et al 2000). Vera et al (2004) found that the difference in El Niño response over southern hemisphere might be mainly driven by atmospheric changes which induce extra tropical SST anomalies. Although ENSO is unquestionably the most important remote forcing, the variability originated by other regional or remote sources cannot be disregarded as well. Many authors have investigated other large scale atmospheric forcing. For example, Barros and Doyle (2002) found a significant relation between the sea surface temperature in the south-western of the Atlantic Ocean in summer and the incoming of humid air into the continent; Silvestri and Vera (2003) related rainfall in autumn and spring with the Antarctic Oscillation.

These results are important for predicting seasonal rainfall in order to plan agricultural activities. The scientific basis of the seasonal climate predictability lies in the fact that slow variations in the earth's boundary conditions (i.e. sea surface temperature or soil wetness) can influence global atmospheric circulation and thus precipitation. As the skill of seasonal numerical prediction models is still limited, it is essential to carry out statistical studies of the probable relationships between some local or remote forcing and rainfall. In this paper an example of seasonal rainfall prediction is presented for the Bermejo River basin.

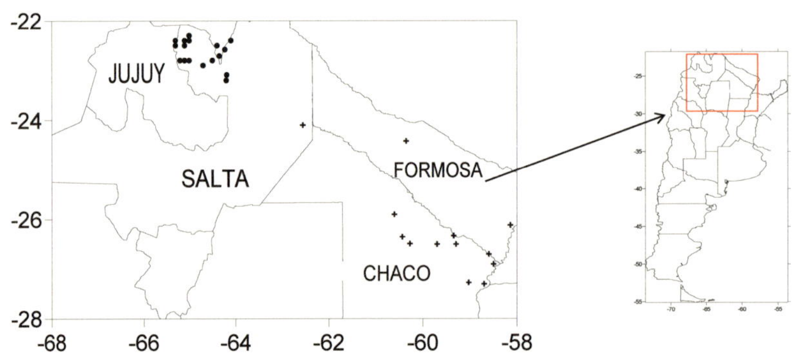

Fig. 1. Stations used in the study. Stations in the Upper (Lower) Bermejo are drawn with a circle (cross).

The objective of this paper is to show the statistical prediction of summer rainfall in the Bermejo River Basin. The paper is organized as follows: Section 2 describes the dataset and the methodology; Section 3 presents the general rainfall features in Bermejo River Basin; Section 4 details the linkage between rainfall over the basin and atmospheric circulation, between rainfall and SST anomaly patterns, and the building of regression models to estimate summer rainfall in the basin; Section 5 presents the main conclusions.

2. Data and methodology

Monthly rainfall data derive from 33 stations from different sources: the National Meteorological Service (SMN), the Secretary of Hydrology of Argentina (SRH), the Regional Commission of the Bermejo River (COREBE) and the Provincial Water Administration of Chaco (APA). The area of study is the Bermejo River Basin, encompassing the Argentinean provinces of Salta, Chaco and Formosa (see Fig 1). Stations have different data records over 1968-2007. All the selected stations have less than 20% of missing monthly rainfall data and their quality has been carefully proved. Precipitation low frequency variability was analyzed using a linear trend method of minimum squares, and statistics significance was tested using a T-Student test. There were also used: monthly sea surface temperature (SST), 500 HPa (G500), 1000 Hpa (G1000) and 200Hpa (G200) geopotential height, zonal (U) and meridional (V) wind at 850 Hpa and 925 Hpa specific humidity (HE) from National Centre of Environmental Prediction (NCEP) reanalysis (Kalnay et al, 1996). Monthly anomalies were determined removing the climatological monthly means from the original values. The analysis in the Bermejo River Basin concentrates on southern summer rainfall. Although it is a small area, some differences were detected all over the basin. Therefore, two mean rainfall series were constructed as the average of monthly precipitation in nineteen stations in the upper Bermejo river basin (UB) during the period 1982-2007 and fourteen stations in the lower and middle Bermejo River Basin (LB) during the period 1968-2007. Such rainfall series are representative of the precipitation over each one of the basin sub-regions (see Fig 1). Different records were considered for each sub-basin regarding data availability.

Simultaneous (summer rainfall with summer circulation variable) and one month lagged (summer rainfall with December circulation variable) correlations were calculated to find the existing relation between summer rainfall and SST and G1000, G500, G200, U and V. Summer rainfall was defined as precipitation accumulated from January to March (hereinafter JFM). The derived correlations allowed defining some predictors which were used to develop a statistical forecast model using the forward stepwise regression method (Wilks 1995). This method retained only the variables, correlated with a 95% significance level. Forward stepwise regression is a model-building technique that finds subsets of predictor variables that most adequately predict responses on a dependent variable by linear regression (Darlington 1990). The basic procedures involve the initial model identification, then predictors are added one-by-one with the remaining candidate predictor that reduces the size of the errors, and this process continues until the errors cannot be significantly reduced. It is important to notice that all the predictors used in the models are variables observed in December (a month prior to rain), averaged in areas where the correlation between these variables and JFM rainfall was high.

To validate the results, it was applied a cross-validation (Wilks 1995) where n-1 years were used for calibration and the remaining year was used to validate the model. This process was repeated several times with a different year as the validation target in each case. This

method is generally strong in the presence of long-term climate variability and is used specially, when the number of data is not so large.
To prove the skill of scheme a contingency table between observed and forecast rainfall was designed using three equiprobable categories labelled below- normal (BS), above normal (AN) and normal (N), referring to the driest, wetter and normal third of cases respectively. Besides, some measures of accuracy were calculated. The most direct and intuitive measure of the accuracy of the categorized forecast (Wilks 1995) is the hit rate or the right proportion. It is the fraction of all the cases whenever the categorical forecast correctly anticipates the subsequent event. The probability of detection is defined as the fraction of those occasions when the forecast events occurred as they had been forecast. The false alarm ratio is the proportion of forecast events that failed to happen. Additionally, empirical estimated and observed rainfall probability functions were calculated using frequency distributions and a chi-square test was used to prove that they did not differ significantly.

3. General rainfall features in Bermejo river basin

To better study the Bermejo River basin the general rainfall features of UB and LB were analyzed. The mean annual rainfall cycle in the common period 1982-2007 was calculated for both sub-basins (Fig 2) and shows an important annual cycle with maximum rainfall in summer (JFM) that increases westwards. Maximum monthly rainfall is greater than 190 mm in January and minimum is around 5 mm in July in UB. The amplitude decrease in LB where minimum rainfall is 21,7 mm in July and maximum reaches only 136 mm in January.
The annual and seasonal rainfall trends (Table 1) were calculated for each sub-basin during the period 1982-2007 in UB and 1968-2007 and 1982-2007 periods in LB. Negative trends were observed as from 1980s, which were significant regarding annual rainfall in UB (-7,93 mm/year) and autumn rainfall in LB (-5,23 mm/year). Seasonal trends are all negative in UB for the period 1982-2007. In the case of LB the annual trend was 1,83 mm/year in 1968-2007 and decreased to -7,09 m/year in 1982-2007, both not significant. In LB, all seasonal trends were positive during 1968-2007, except southern winter trend (-1,44 mm/year). However, when considering the 1982-2007 period, all seasonal trends became negative, although only autumn trend resulted significant. Therefore, there is evidence that rainfall tended to decrease from the1980s. In LB, where data are available, a relevant change in the behaviour of trends could be detected. It is important to mention that rainfall has increased all over the northeast and central Argentina during the last century (Barros et al, 2008 and 2000; Liebmann et al, 2004; Gonzalez et al, 2005) but there is some evidence that these trends could be reversing in the Chaco plain region of Argentina (Flores and Gonzalez, 2009) in recent times. So, it would be great to determine the possible causes which are related to these observed changes. For example, LechuSA programme (Land Ecosystem change utility for South America, http://lechusa.unsl.edu.ar/) (Baldi et al 2008) is an enterprise whose main objective is to understand the changes of ecosystems in South America, specially, those derived from changes in land uses. They identified "hotspots" with satellite information which is representative of intense changes. One of these "hotspots" is located in the Bermejo River Basin due to the deforestation and intensification of agricultural activities. Also, Minetti et al (2010) have shown an upward trend in the occurrence of droughts from 2003 in Argentina, including the area analyzed in this paper.

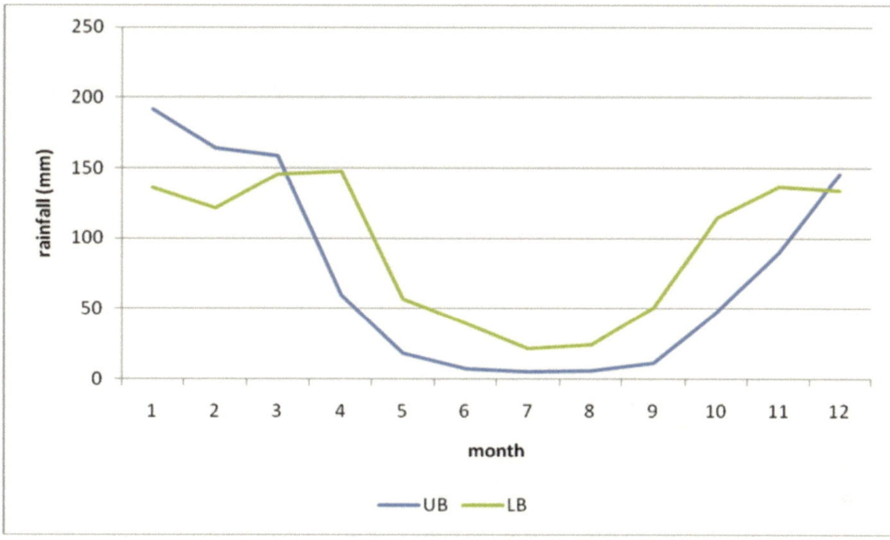

Fig. 2. Mean annual cycle of rainfall (mm) in both sub-basins (UB blue line, LB green line) defined in the text for the period 1982-2007.

Rainfall trend in LB (1968-2007)		
	trend (mm/year)	R
ANNUAL	1,83	0,11
JFM	0,81	0,10
AMJ	0,07	0,01
JAS	**-1,44**	0,38
OND	2,39	0,30

Rainfall trend in UB (1982-2007)		
	trend (mm/year)	r
ANNUAL	**-7,93**	0,44
JFM	-3,39	0,27
AMJ	-0,34	0,30
JAS	-0,28	0,18
OND	-2,67	0,34
Rainfall trend in LB (1982-2007)		
	trend (mm/year)	R
ANNUAL	-7,09	-0,30
JFM	-1,24	-0,10
AMJ	**-5,23**	-0,43
JAS	-1,77	-0,33
OND	1,15	0,09

Table 1. Annual and seasonal rainfall trends (mm/year) and the correlation coefficient (r) in the availability record (left panel) and from the 1980s (right panel) for UB (upper panel) and LB (bottom panel). Significant correlations at 95% confidence level are enhanced in italics. The availability record for UB is 1982-2007, so only one period is detailed.

Although there is some controversy about the actual causes of land use changes and the relationship with observed rainfall trends, some authors have analyzed the impact of land cover over climate (Lee and Berbery, 2010; Jin et al, 2000; Mabuchi et al, 2005, among others). In particular, Lee (2010) and Lee and Berbery (2010) studied these impacts in the La Plata Basin, where the Bermejo River Basin is included. They proved that areas where croplands replace forests would experience an increase in albedo, a reduction of surface friction and sensible heat. The effect on sensible heat seems to dominate and leads to a reduction in convective instability and the stronger low level winds due to reduced friction enhance moisture advection out of the basin. The two effects, increased stability and reduced moisture flux, result in a reduction of precipitation.

4. Summer rainfall prediction

As the maximum rainfall season was summer (JFM), this period will be the subject of study from now onwards. In order to create a prediction scheme for JFM rainfall, simultaneous (JFM rainfall with JMF circulation variable) and one month (JFM rainfall with December circulation variable) lagged correlations were calculated to find the existing relationship between summer rainfall and SST, G1000, G500, G200, U and V in each sub-basin. The 1982-2007 and 1968-2007 data periods were used for UB and LB, respectively. Therefore, the correlation required for significance at the 95% confidence level was 0,4 for UB and 0,32 for LB. The areas with high correlation between JFM rainfall and the previous month circulation features (circulation variable in December) were used to define some predictors which will be detailed in Table 2. Only in some cases the simultaneous correlation maps will be described as well in order to determinate if the signal lasts until the rain. However, predictors to carry out the regression scheme for JFM rainfall, will be SST or circulation variables in December; they will be carefully selected, based on statistical significance and physical reasoning.

4.1 Summer rainfall prediction in LB

The first stage of the model development was the analysis of the statistical association between JFM rainfall in LB and global SST in December. The correlation map (figure not shown) shows significant positive correlation, only in an area along the coast of Southern Brazil, between (25ºS-40ºS; 290ºE-340ºE) (AS), indicating that warm sea is associated to rainfall in LB.

The correlation maps between JFM rainfall in LB and G200 (figure not shown), G500 (Fig 3) and G1000 (Fig 4) in December show an annular-like pattern (SAM, Southern Annular Mode, Thompson and Wallace 2000) at high latitudes combined with wave-like pattern at middle latitudes. In general, a measure of SAM is the Antarctic Oscillation index (AAO), that is defined as the leading principal component of 850 hPa geopotential height anomalies south of 20ºS (Thompson and Wallace 2000). The SAM is a nearly annular pattern with a large low pressure anomaly centred on the South Pole and a ring of high pressure anomalies at mid-latitudes. This feature increases zonal winds, decreases heat exchange between poles and mid-latitudes and so modifies storm tracks. The main feature in Figs 3 and 4 indicates that the presence of this positive SAM phase in December is associated with rainfall greater than normal in JFM in LB. Indeed, the correlation between AAO in December and JFM rainfall in LB is 0,39, as it is detailed in Table 2. Some authors have shown that SAM might have influence on rainfall variability in some regions of the Southern Hemisphere. For example, Zheng and Frederiksen (2006), show that this signal affects summer rainfall variability in the New Zealand; Reason and Rouault (2005) have shown that wetter (drier) winters in western South Africa occur during negative (positive) SAM phase; Meneghini et al (2007) found a significant inverse relationship between SAM and rainfall in southern Australia with a significant in-phase relationship in northern Australia. Some results have been published relating SAM and rainfall in South America: Silvestri and Vera (2003) and Reboita et al (2009) show such influence on precipitation interannual variability in southern Brazil, particularly during spring.

Associated to this SAM pattern, a regional anticyclonic anomaly is positioned in central and southern Argentina in middle and high levels and so two predictors could be defined: G5-1 (G2-1), as the mean G500 (G200, figure not shown) in (35°S-45°S;95°W-70°W) in December. In low levels (G1000), a significant negative correlation in observed in southern Brazil and over the basin indicating that JFM rainfall is related to the weaken Atlantic High in December. This result agrees with Gonzalez (2010) when they analyze summer rainfall in the Argentinean Chaco rainfall. Fig 4 shows the correlation field between JFM rainfall en LB and G1000 in December. There, the predictor G10-1 was defined as the mean G1000 (20°S-30°S; 55°W-40°W) in December. Associated with SAM, G2-A, G5-A and G10-A were defined as the mean G200, G500 and G1000 respectively in the area (75°S-90°S; 0°E-360°E) in December.

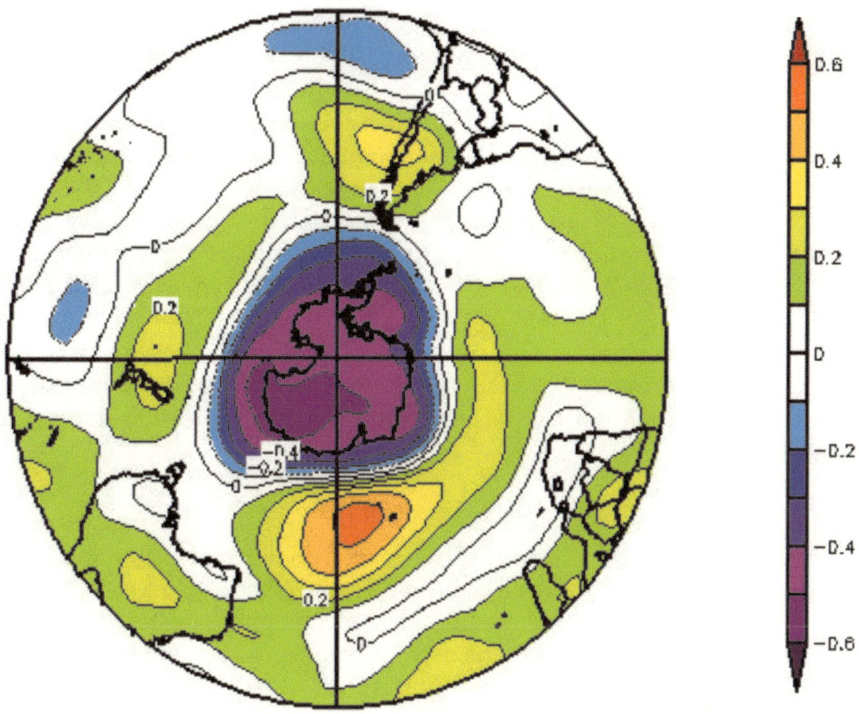

Fig. 3. Correlation map between JFM rainfall in LB and G500 in December. Dash lines are negative values. Correlation greater tan 0,32 are significant at 95% confidence level.

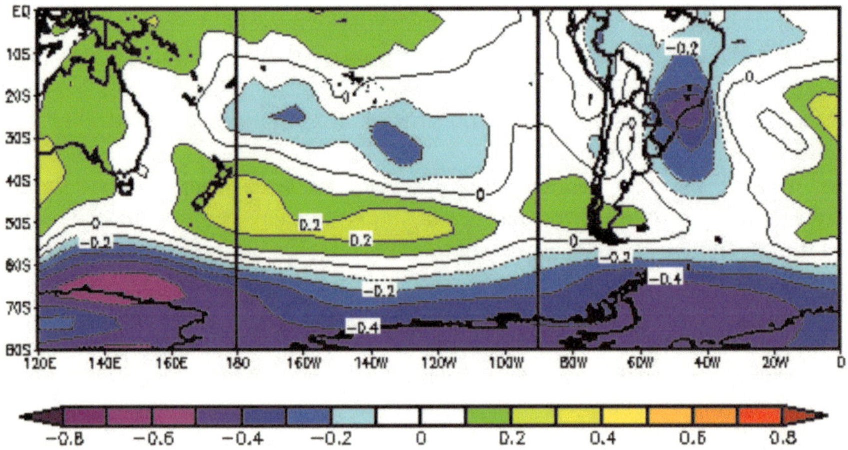

Fig. 4. Correlation map between JFM rainfall in LB and G1000 in December. Dash lines are negative values. Correlation greater tan 0,32 are significant at 95% confidence level.

Simultaneous correlation maps between JFM rainfall in LB and JFM U (Fig 5), V (Fig 6) and HE (Fig 7) were constructed. They show that rainfall is associated with a pronounced low level jet located in Bolivia and Paraguay and north of Argentina that is depicted by north-western winds over LB (figures 5 and 6) and with high values of HE in north-eastern Argentina and southern Brazil (Fig 7). Although these variables are relevant in relation to enhanced precipitation derived from the moist air advection, they cannot be regarded in the regression model because all these patterns do not remain when one month lag (JFM rainfall and U,V and HE in December) is considered for correlation (figures not shown).
The correlation between JFM rainfall in LB and the predictors in December defined in the text are detailed in Table 2. A multiple linear regression analysis was carried out using the predictors best correlated with JMF rainfall in LB and with no significant correlation among them in order to avoid a multicolinearity problem. In summary, the predictors put into the regression model were: AS, G2-1, G2-A and G10-1. The forward stepwise method selected the predictors: G2-A and G10-1. The equation of linear regression forecast model was formulated as follows:

$$R_{LB} = -5{,}54\ G2\text{-}A - 5{,}71\ G10\text{-}1 + 7160{,}8$$

where R_{LB} (in mm) is the estimated JFM rainfall in LB and G2-A and G10-1 are the predictors (in m). The R-square was 0,491 and so, the percentage of variance explained by the model was 49%. A commonly used measure of strength of the regression is the F-ratio, defined as the relationship between the mean square regression and the mean square error (Wilks, 1995). It is well-known that the F-ratio is high when a strong relationship between R and the predictors produce large mean square regression and small mean square error. As the residuals of the regression are independent and follow a normal distribution, under the null hypothesis of no lineal regression, the F-ratio is 17,8 with a p-value of 0,00001, and therefore, the regression model provides reasonably forecast with 95% confidence. The method only retained the predictors G2-A and G10-1 from the four candidate predictors, indicating that the two main factors that influence summer precipitation in that area are: the

SAM phase and the weaken Atlantic High. The observed and forecast rainfall time series derived from cross-validation are shown in Fig 8. The correlation between them (0,6) is significant at the 95% confidence level.

	LB	UB
AAO	**0,39**	0,15
EN34	-0,01	**-0,43**
PE	-0,02	**-0,47**
PS	-0,08	**0,48**
IT	0,23	**-0,52**
AS	**0,32**	0,24
g5-1	**0,33**	0,38
g5-2	0,16	**0,46**
g5-3	**-0,42**	-0,38
g5-4	0,09	0,31
g5-5	0,1	-0,16
g5-A	**-0,5**	-0,1
g10-1	**-0,41**	-0,04
g10-2	0,09	**0,45**
g10-3	**-0,39**	-0,37
g10-4	0,04	0,31
g10-5	-0,05	-0,21
g10-A	**-0,46**	-0,18
g2-1	**0,39**	**0,41**
g2-2	0,3	**0,42**
g2-3	**-0,38**	**-0,41**
g2-4	0,2	0,3
g2-5	0,24	-0,14
g2-A	**-0,51**	-0,13
g2-7	0,05	0,13
UP	-0,06	**-0,4**
UA	0,06	0,24

Table 2. Correlations between the predictors defined in the text for December and the mean JFM rainfall series in UB, and LB. Enhanced in italics are values at the 95% confidence level (greater than 0,32 for LB and 0,4 for UB).

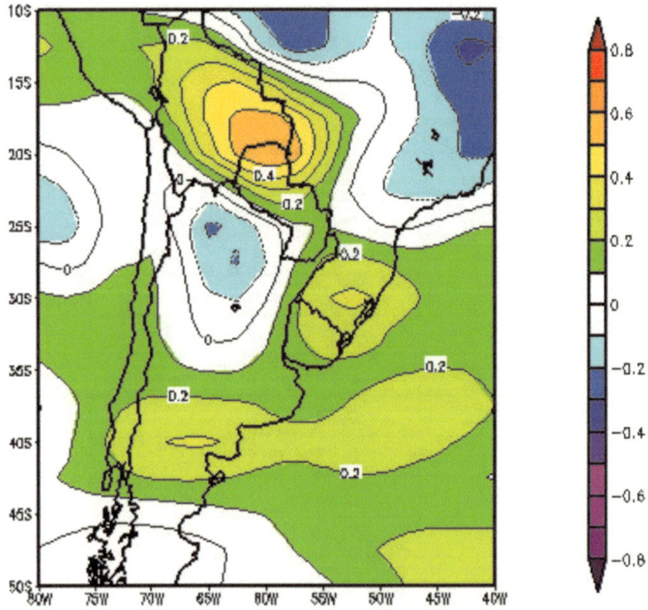

Fig. 5. Correlation map between JFM rainfall in LB and U in JFM. Dash lines are negative values. Correlation greater tan 0,32 are significant at 95% confidence level.

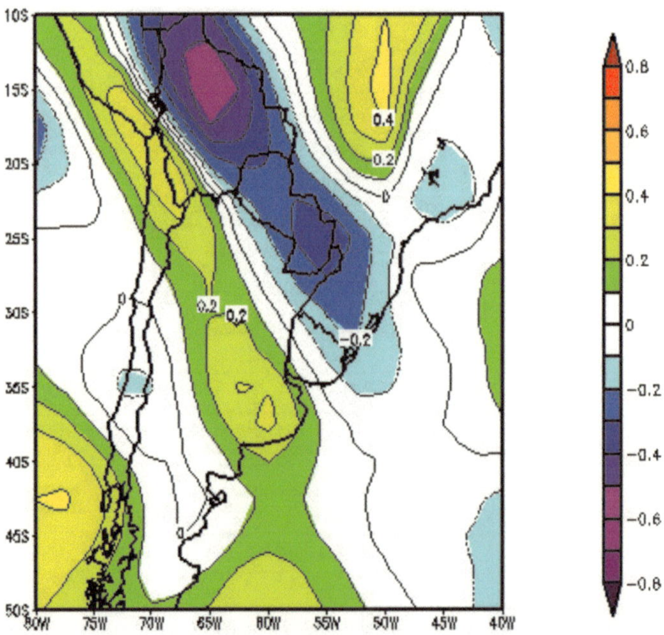

Fig. 6. Correlation map between JFM rainfall in LB and V in JFM. Dash lines are negative values. Correlation greater tan 0,32 are significant at 95% confidence level.

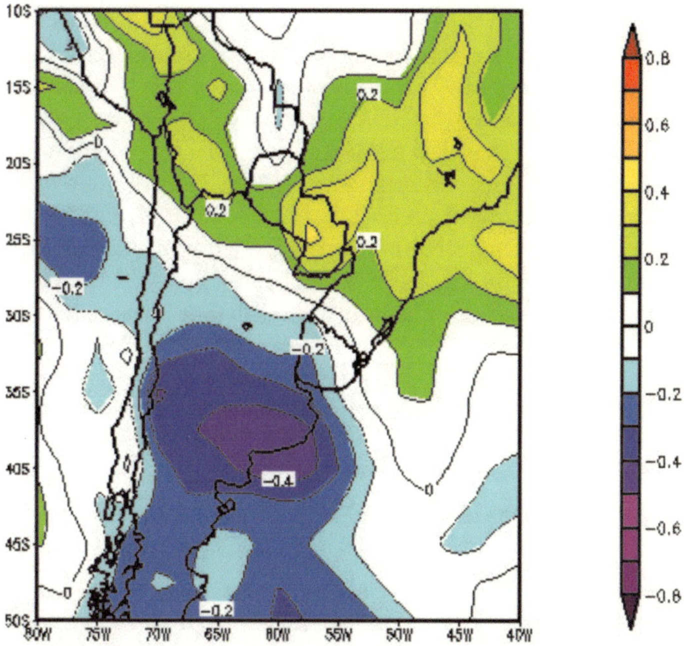

Fig. 7. Correlation map between JFM rainfall in LB and HE in JFM. Dash lines are negative values. Correlation greater tan 0,32 are significant at 95% confidence level.

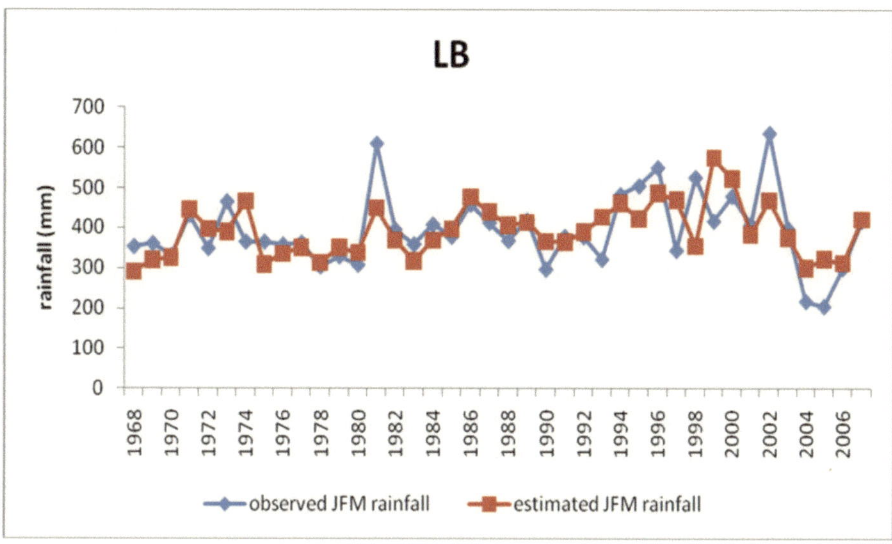

Fig. 8. Observed and estimated with a crossvalidation scheeme JFM rainfall in LB for the period 1968-2007.

4.2 Summer rainfall prediction in UB

Fig 9 shows the linear correlation between JFM rainfall in UB and global SST in December. Negative correlation values along the tropical band in the Pacific Ocean (PE, 5ºS-5ºN; 140ºW-130ºW) were observed and they are associated with the El Niño-Southern Oscillation (ENSO) phenomenon, indicating that rainfall is mainly related to the cold phase of ENSO. Indeed, correlation between JFM rainfall in UB and December SST in the region El Niño 3.4 is -0,43, significant at the 95% confidence level (Table 2). Another area with positive correlation was detected in southwest Pacific Ocean (PS, 35ºS-52ºS; 170ºW-160ºW). Besides, a region with significant negative correlation was positioned in the Indian Ocean (IT, 0º-10ºS; 60ºE-70ºE). Some authors have pointed out relationships between rainfall and SST in the Indian Ocean (Zheng and Frederiksen 2006 in New Zeland, Reason 2001 in South Africa, Gissila 2004 in Ethiopia, Gonzalez and Vera 2009 and Gonzalez et al 2010, in South America). These significant correlations persist when simultaneous correlation is performed (figures not shown). In fact, Mo (2000) described tropical Indian Ocean SST anomalies on quasi-biennial timescales in association with a "Pacific South American Pattern" characterized by a wave train emanating from the tropical western Pacific poleward. This wave train reaches South America, enters Argentina south 40ºS where the Andes Mountain is lower and displaces towards the northeast of Argentina as polar fronts, producing precipitation in central and northern Argentina.

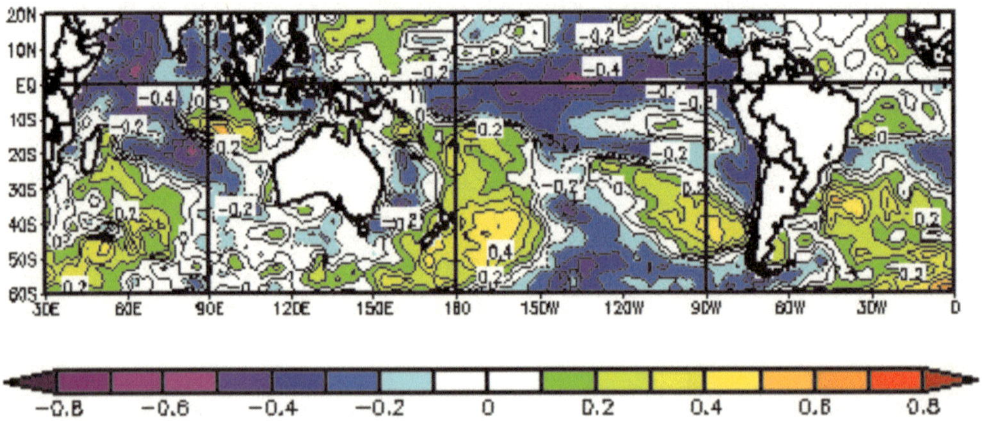

Fig. 9. Correlation map between JFM rainfall in UB and SST in December. Dash lines are negative values. Correlation greater tan 0,4 are significant at 95% confidence level.

The correlation maps between G500 (Fig 10), G1000 and G200 (figures not shown) anomalies for December and JFM rainfall in UB, show the same pattern: an anticyclonic anomaly all over the central and southern part of Argentina and an intensification of low pressure south of 60ºS. This pattern is very similar to the one described for LB but SAM is not as well defined. Therefore, two predictors could be defined: G5-2 as the mean G500 (45ºS-55ºS; 85ºW-60ºW) and G5-3 in (70ºS-75ºS; 130ºW-90ºW) in December. Similar predictors have been defined in G1000 (G10-2 and G10-3) and G200 (G2-2 and G2-3) for the same regions in December. The correlations map between JFM rainfall and V in December doesn`t show significant values in UB. However, the correlation map with U in December (Fig 11) shows an area of negative correlation, south of 35ºS, indicating that weaken western winds in

December are associated with enhanced rainfall in JFM and so, the predictor UP was defined as the mean zonal wind in (40ºS-48ºS;75ºW-60ºW). The simultaneous correlation map with HE only shows significant values in north-eastern Argentina (Fig 12). These correlations decreased and became non-significant when a one month lag was used, so no predictors derived from this variable could be defined.

The predictors IT, PS, G5-2 and G2-3 were entered in the forward stepwise regression model because they were significant correlated to UB rainfall and they were independent each other and the resulting equation is:

$$\mathbf{R_{UB}} = -162{,}86\ \mathrm{IT} + 74{,}55\ \mathrm{PS} + 4130$$

$\mathbf{R_{UB}}$ is expressed in mm, IT and PS in °C. The F-ratio is 11,67 with p< 0,00032. The R-square is 0,504 and therefore, the model explains the 50% of the variance of JFM rainfall in UB. The crossvalidation method was applied and the correlation between observed and forecast series was 0,49 (Fig 13). This result shows that in UB, SST is the main variable which affects summer rainfall.

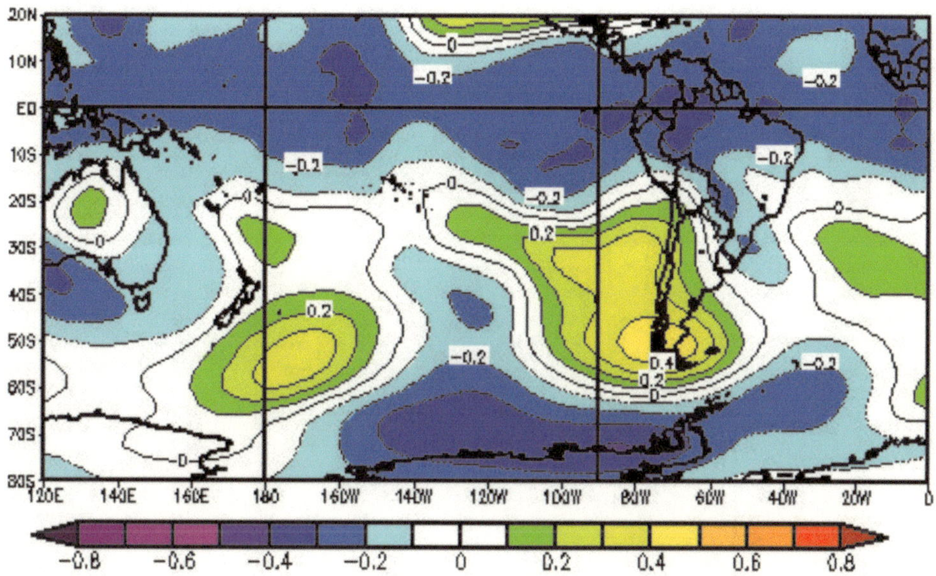

Fig. 10. Correlation map between JFM rainfall in UB and G500 in December. Dash lines are negative values. Correlation greater tan 0,4 are significant at 95% confidence level.

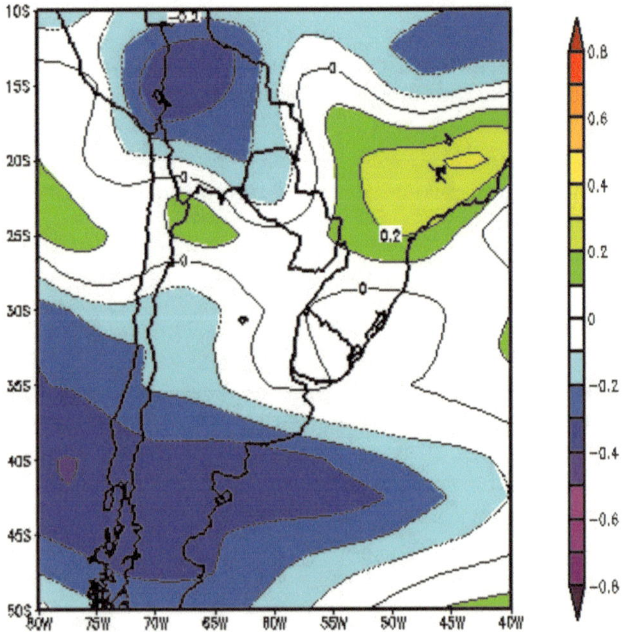

Fig. 11. Correlation map between JFM rainfall in UB and U in December. Dash lines are negative values. Correlation greater tan 0,4 are significant at 95% confidence level.

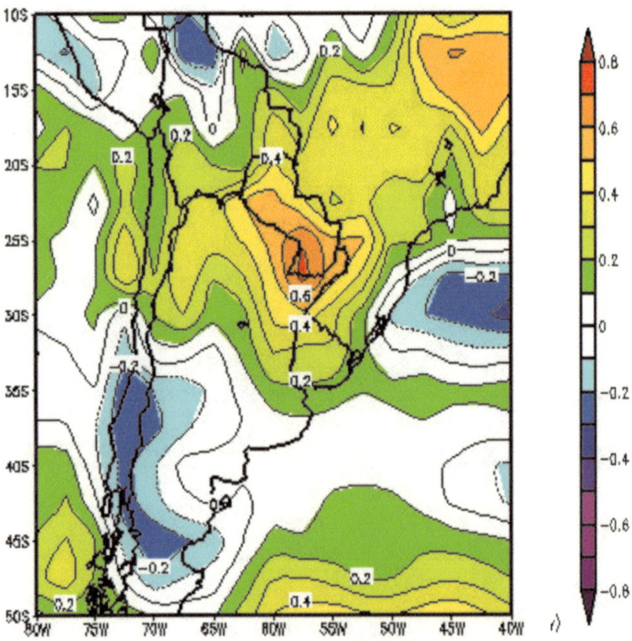

Fig. 12. Correlation map between JFM rainfall in UB and HE in JFM. Dash lines are negative values. Correlation greater tan 0,4 are significant at 95% confidence level.

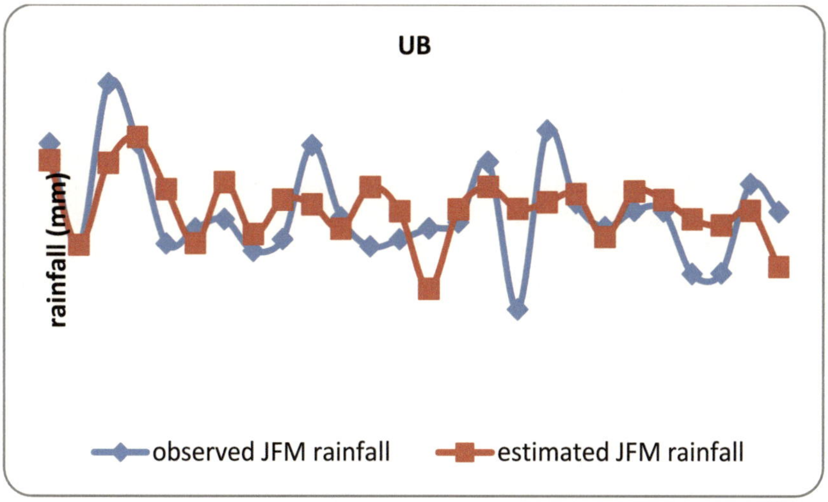

Fig. 13. Observed and estimated with a crossvalidation scheeme, JFM rainfall in UB for the period 1982-2007.

4.3 Forecast skill

A measure to evaluate how well the predictions match the observed rainfall values is based on the contingency table. The observed and estimated values are assigned to one of three equiprobable categories labelled below- normal (BS), above normal (AN) and normal (N), referring to the driest, wetter and normal third of cases respectively. A table comparing the forecast and observed rainfall for the three sub-basins were constructed (Tables 3 and 4). The correct predictions are shown in the diagonal cells from top left to bottom right. A chi square test was used to prove that both tables are significantly different from random at the 95% confidence level. Table 3 reveals that most cases have been correctly classified in LB. However, table 4 shows that many times below and above observed rainfall have been estimated as normal in UB, revealing a tendency to underestimate the extreme cases.

		FORECAST CATEGORIES		
		BN	N	AN
OBSERVED CATEGORIES	BN	22,59	5	5
	N	10	20	2,5
	AN	0	7,5	27,5

Table 3. Contingency table for LB (in % of the total cases).

		FORECAST CATEGORIES		
		BN	N	AN
OBSERVED CATEGORIES	BN	15,4	11,5	7,7
	N	19,2	7,7	3,8
	AN	0	11,5	23,1

Table 4. Contingency table for UB (in % of the total cases).

To evaluate the accuracy of the categorized forecast, the hit rate (H), the probability of detection (POD) and the false alarm relation (FAR) were calculated for each category. Table 5 shows these accuracy measures for the events "BN" and "AN". They were calculated collapsing the 3x3 contingency table into two 2x2 tables. Each one was constructed by considering the "forecast event" (BN and AN) in distinction to the complementary "non forecast event" (non-BN or non-AN). The values in Table 5 indicate that the method detects some cases of extreme rainfall, improving the climatology. Results in LB are better than in UB. The probability to detect above normal rainfall events is in general, better than the probability to detect below normal rainfall ones. The probability to give a false alarm in a below normal rainfall event is greater than in the above normal cases.

	LB		UB	
	BN	**AN**	**BN**	**AN**
H	0,8	0,85	0,62	0,77
POD	0,69	0,79	0,44	0,67
FAR	0,31	0,21	0,56	0,33

Table 5. Measures of accuracy for above-normal rainfall events (AN) and for below normal rainfall events (BN) for each one of the sub-basins (LB and UB).

In order to convert the individual estimations in a probabilistic forecast, the accumulated frequencies were calculated for the observed and forecast rainfall values and the empiric probability functions were drawn for each sub-basin (Fig 14 and 15). A chi square test was used for each pair of probability functions and in all cases they resulted significantly similar each other at the 95% confidence level. These figures reveal that, in both LB and UB, the derived prediction schemes tend to underestimate the extremes. However in LB sub-basin, cases of precipitation greater than 300 mm and lower than 400mm, were well detected.

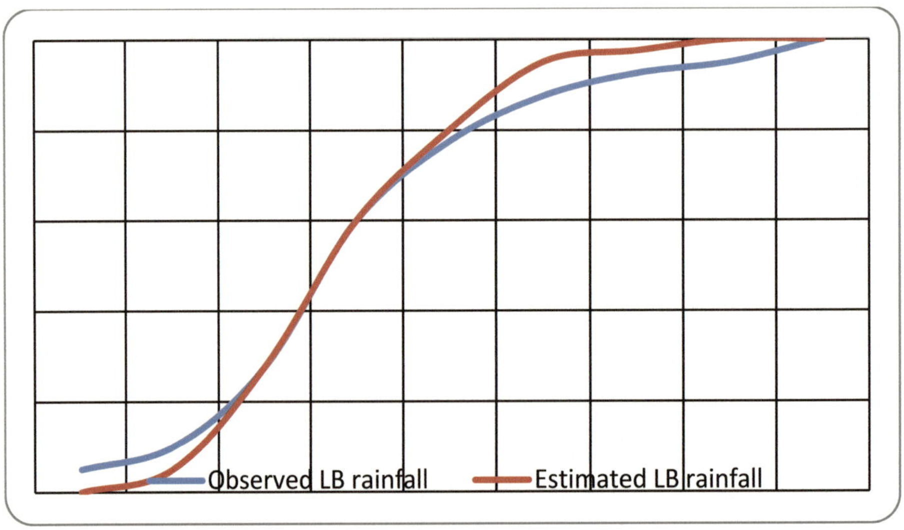

Fig. 14. Empiric probability function for observed (blue line) and estimated (red line) JFM rainfall in LB (in mm).

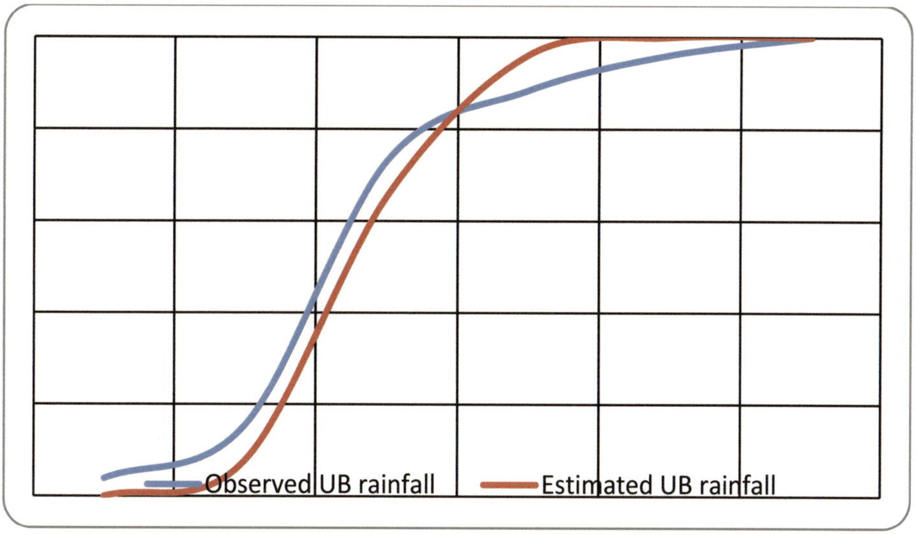

Fig. 15. Empiric probability function for observed (blue line) and estimated (red line) JFM rainfall in UB (in mm).

5. Conclusions

The Bermejo River Basin has experienced a relevant annual rainfall decrease since 1980, especially in summer and autumn. There is a large annual cycle with maximum in summer, all over the region but it is larger towards the west. As the maximum rainfall season was summer (JFM), this period was used to estimate rainfall using predictors defined in December, for each sub-basin (LB and UB). The two main factors that influence summer precipitation in LB were the SAM phase and the weaken Atlantic High. The correlation between observed and forecast rainfall time series derived from cross-validation was 0,6 and the linear regression model explained the 49% of the variance of JFM rainfall in LB. However, the summer rainfall in UB depended mainly of the Pacific and Indian Ocean sea surface temperatures. The final model explained the 50% of the variance of JFM rainfall in UB; the correlation between observed and forecast series was 0,49.
The efficiency of the method was proved by calculating some statistics like the hit rate, the probability of detection and the false alarm ratio. Results in LB are better than in UB. The probability of above normal rainfall events is better than the probability to detect below normal rainfall ones. The probability to give a false alarm in a below normal rainfall event is greater than in the above normal cases. Additionally, the probability functions resulting from estimated and observed JFM rainfall was similar at the 95% confidence level and revealed that the method underestimates the most extreme cases.
These results are promising and encourage further work in order to examine new techniques to better estimate rainfall, especially the extremes, and to investigate other predictors which could affect precipitation in summer.

6. Acknowledgements

Rainfall data were provided by the National Meteorological Service (SMN), the Secretary of Hydrology of Argentina (SRH) and the Provincial Water Administration of Chaco (APA) and the Regional Commission of Bermejo River (COREBE). Images from figures 3 to 7, 9 to 12 were provided by the NOAA/ESRL Physical Sciences Division, Boulder Colorado from their web site: http://www.cdc.noaa.gov. This research was supported by UBACyT CC02, UBACyT 1028 (2011-2014) and CONICET PIP 112-200801-00195.

7. References

Baldi, G.; Nosetto, M. & Jobbagy, E. (2008). The effect of tree plantations on south American ecosystems functioning. *Ambiencia,* Vol. 4, pp. 23-34, ISSN 1808-0251.

Barros, V.; Castañeda, M.E. & Doyle, M. (2000). Precipitation trends in Southern South America, east of the Andes: An indication of climate variability, In: *Southern Hemisphere Paleo-and Neoclimates: Key Sites, Methods, Data and Models,* Peter Smolka and Wolfgang Volkheimer, pp 187-208, Springer.

Barros, V.; Doyle, M.; González, M.H.; Camilloni, I.; Bejarán, R. and Caffera, M. (2002). Revision of the south american monsoon system and climate in subtropical South America south of 20°S. *Meteorologica,* Vol. 27, pp. 33-58, ISSN 1325 187X.

Barros, V. and Doyle, M. (2002). Mid-summer low level circulation and precipitation in subtropical South America and related sea surface temperature anomalies in the South Atlantic. *J Climate,* Vol. 15, pp. 3394-3410, ISSN 0894 8755.

Barros, V.; Doyle, M. and Camilloni, I. (2008). Precipitation trends in southeastern South America: relationship with ENSO phases and the low-level circulation. *Theoretical and Appl Climatology,* Vol. 93 1-2, pp. 19-33, ISSN 0177-798X

Darlington, R.B. (1990). *Regression and linear models.* McGraw-Hill, ISBN 0070153728, New York.

Flores, O.K. and González, M.H. (2009). Análisis de la precipitación en la llanura chaqueña argentina y su relación con el comportamiento de la circulación atmosférica y las temperaturas de la superficie del mar. *Proceeding of CONGREMET X - CLIMET XIII,* ISBN 978-987-22411-1-7, Buenos Aires Argentina, October 2009.

Gissila, T.; Black, E.; Grime, D.I.F. and Slingo, J.M. (2004). Seasonal forecasting of the Ethiopian summer rains. *Int J Climatol,* Vol. 24, pp. 1345-1358, ISSN 0899-8418.

González, M.H. and Barros, V. (1998). The relation between tropical convection in South America and the end of the dry period in subtropical Argentina. *Int J Climatol,* Vol. 18 No. 15, pp.1669-1685, ISSN 0899-8418.

González, M.H. ; Castañeda, M.E. and Texeira Neri, J. (2005). Evolución de la precipitación en el noreste de Argentina y sur de Brasil. *Proceeding Congremet IX* , ISBN 987-22411-0-4, Buenos Aires Argentina, October 2005.

González, M.H. and Vera, C.S. (2010). On the interannual winter rainfall variability in Southern Andes. *Int J Climatol,* Vol. 30, pp. 643-657, ISSN 0899-8418.

González, M.H., Skansi, M.M. and Losano, F. (2010). A statistical study of seasonal winter rainfall prediction in the Comahue region (Argentine), *ATMOSFERA,* Vol. 23, No. 3, pp. 277-294, ISSN 0187-6236.

González, M.H. (2010). Los cambios observados en la precipitación en el Chaco Argentino. *Proceeding II Congreso venezolano de Agrometeorología,* El Tigre Venezuela, May 2010.

Grimm, A.; Barros, V. and Doyle, M. (2000). Climate variability in Southern South America associated with El Niño and La Niña events. *J Climate,* Vol. 13, pp. 35-58, ISSN 0894 8755.

Jin, J.; Lu, S.; Li, S. and Miller, N.L. (2000). Impact of Land Use Change on the Local Climate over the Tibetan Plateau. *Advances in Meteorology,* doi:10.1155/2010/837480, ISSN 1687-9309.

Lee, S.J. and Berbery, E.H. (2010). On the effects of land cover change on the climate of La Plata Basin. *J Hydrometeor,* submitted, ISSN 1525-755X.

Liebmann, B.; Vera, C.; Carvalho, L.; Camilloni, I.; Hoerling, M.; Allured, D.; Barros, V; Báez, J. and Bidegain, M. (2004). An Observed Trend in Central South American Precipitation. *J Climate,* Vol. 17, No. 22, pp. 4357-4367, ISSN 0894 8755.

Kalnay, E; Kanamitsu, M.; Kistler, R.; Collins, W.; Deaven, D.; Gandin, L.; Iredell, M.; Saha, S.; White, G.; Woollen, J.; Zhu, I.; Chelliah, M.; Ebisuzaki, W.; Higgings, W.; Janowiak, J.; Mo, K.C.; Ropelewski, C.; Wang, J.; Leetmaa, A.; Reynolds, R.; Jenne, R. and Joseph, D. (1996). The NCEP/NCAR Reanalysis 40 years- project. *Bull Amer Meteor Soc,* Vol. 77, pp. 437-471, ISSN 1520-0477.

Kiladis, G. and Diaz, H. (1989). Global Climatic Anomalies associated with extremes in the Southern Oscillation. *J Climate,* Vol. 2, pp. 1069-1090, ISSN 0894 8755.

Lenters, J.D. and Cook, K.H. (1997). On the origin of Bolivian High and related circulation feature of the South American Climate. *J Atmos Sci,* Vol. 54, pp. 656-677, ISSN 0022-4928.

Mabuchi, K.; Sato, Y. and Kida, H. (2005). Climatic Impact of Vegetation Change in the Asian Tropical Region. Part I: Case of the Northern Hemisphere Summer. *J Climate,* Vol. 18, pp. 410-428, ISSN 0894 8755.

Minetti, J.L.; Vargas, W.M.; Poblete, A.G.; De la Zerda, L.R. and Acuña, L.R. (2010). Regional rought in Southern South America. *Theor Appl Climatol.* DOI 10.1007/s00704-010-0271-1, ISSN 0177-798X.

Mo, K.C. (2000). Relationships between low frequency variability in the Southern Hemisphere and sea surface temperature anomalies. *J Climate,* Vol. 13, pp. 3599-3610, ISSN 0894 8755.

Reason, C. (2001). Subtropical Indian Ocean SST dipole events and Southern Africa rainfall. *Geophys Res Lett,* Vol. 28, pp. 2225-2227, ISSN 0094-8276.

Reason , C. and Rouault, M. (2005). Links between the Antartic Oscillation and winter rainfall over western South Africa. *Geophys Res Lett,* Vol. 32, DOI: 10.1029/2005GL022419, ISSN 0094-8276.

Reboita, M.S.; Gan, M.A.; Da Rocha, R.P. and Ambrizzi, T. (2010) Regimes de Precipitacao na America Do Sul. *Revista Brasilera de Meteorologia,* Vol. 25, No. 2, pp. 185-204, ISSN 0102-7786.

Ropelewski, C. and Halpert, M. (1987). Global and Regional scale precipitation patterns associated with El Niño. *Mon Wea Rev,* Vol. 110, pp. 1606-1626, ISSN 0027-0644.

Silvestri, G. and Vera, C.S. (2003). Antarctic Oscillation signal on precipitation anomalies over southeastern South America. *Geophys Res Lett,* Vol. 30, No. 21, pp. 21-15, ISSN 0094-8276.

Thompson, D.W. and Wallace, J.M. (2000). Annular modes in the extratropical circulation. Part I: month-to-month variability. J *Climate,* Vol. 13, pp. 1000-1016, ISSN 0894 8755.

Vera, C.S.; Silvestri, G.; Barros, V. and Carril, A. (2004). Differences in El Niño response in Southern Hemisphere. *J Climate,* Vol. 17, No. 9, pp. 1741-1753, ISSN 0894 8755.

Vera, C.S.; Higgins, W.; Amador, J.; Ambrizzi, T.; Garreaud, R.; Gochis, D.; Gutzler, D.; Lettenmaier, D.; Marengo, J.; Mechoso, C.R.; Nogues-Paegle, J.; Silva Dias, P.L. and

Zhang, C. (2006). Toward a unified view of the American Monsoon Systems. *J Climate,* Vol. 19, No. 20, pp. 4977-5000, ISSN 0894 8755.

Wang , M. and Paegle, J. (1996). Impact of analysis uncertainty upon regional atmospheric moisture flux. *J Geophys Res,* Vol. 101, pp. 7291–7303, ISSN 0148-0227.

Wilks, D.S. (1995) *Statistical methods in the atmospheric sciences (An introduction),* Academic Press, ISBN 0127519653, California USA.

Zheng, X. and Frederiksen, C. (2006). A study of predictable patterns for seasonal forecasting of New Zealand rainfall. *J Climate,* Vol. 19, pp. 3320-3333, ISSN 0894 8755.

Part 5

Adaptation Issues and Climate Variability and Change

8

Climate Change Adaptation in Developing Countries: Beyond Rhetoric

Aondover Tarhule
Department of Geography and Environmental Sustainability
University of Oklahoma,
Oklahoma,
USA

1. Introduction

Developing countries face unique vulnerability and adaptation challenges related to climate variability and change as a result of being, on the one hand, more exposed and sensitive, and on the other hand, having less adaptive capacity for dealing with it (Yohe *et al.*, 2006; UNFCCC, 2007; World Bank, 2010). Also widely accepted is the urgent need for adaptation to combat what the Inter Governmental Panel on Climate Change (IPCC) sees as the most likely climate change impacts in the developing world. These include;

"reduced crop yields in tropical areas leading to increased risk of hunger, spread of climate sensitive diseases such as malaria, and an increased risk of extinction of 20-30 percent of all plant and animal species." The report continues, "by 2020, up to 250 million people in Africa could be exposed to greater risk of water stress. Over the course of this century, millions of people living in the catchment areas of the Himalayas and Andes face increased risk of floods as glaciers retreat followed by drought and water scarcity as the once extensive glaciers on these mountain ranges disappear. Sea level rise will lead to inundation of coasts worldwide ...people living with the constant threat of tropical cyclones now face increased severity and possibly increased frequency of these events with all associated risks to life and livelihoods" (UNFCC, 2007, p. 5).

In the face of such dire consequences, developing countries especially and international organizations, led most prominently by the UNFCCC (United Nations Framework Convention on Climate Change; see http://unfccc.int/2860.php), have been hard at work in an attempt to design feasible climate change adaptation policies and actions. The nexus of these efforts between the developing countries and the UNFCCC is a mechanism called NAPA (National Adaptation Programmes of Actions), which were instituted by Decision 5 at the 7th Conference of Parties (CoP) held in Marrakesh, Morocco, in 2001.

These efforts have galvanized political action and provided much needed guidance to developing countries on how to plan for adaptation to predicted climate change impacts. Even more importantly, they have also provided funding mechanisms and commitments from developed countries to help the least developed countries (UNFCC Article 4/3-5) implement adaptation strategies.

Despite such seemingly impressive progress, a number of critical issues related to adaptation remain to be resolved if we are to move beyond the rhetoric of broad scale policy

to implementation at the scale at which most ordinary people experience the impacts of climate change and climate variability. Using primarily case examples from Africa but also elsewhere in the developing world, this chapter addresses a number of such issues, including the role and place of adaptation in development policy, the question of what to adapt to, as well as the specifics of how to implement adaptation.
The chapter is laid out as follows. Section 1 provides a definition and discussion of relevant terms; Section 2 focuses on the above identified critical issues, and Section 3 lists further actions and steps needed to implement adaptation effectively in developing countries.

2. Definition and elaboration of terms

Climate change adaptation, defined as "initiatives and measures to reduce the vulnerability of natural and human systems against actual or expected climate change effects" (IPCC, 2007, p. 76), represents the right end of the spectrum of individual and societal responses to climate change. At the left end of that spectrum is mitigation, which includes policies and activities designed either to reduce the entry of additional Green House Gases (GHGs) into the atmosphere or reduce the concentration of existing atmospheric GHGs. Despite present ubiquity in the scholarly literature and policy domain, adaptation has not always been popular or even central in the climate change discourse. As noted by Ayers and Dodman (2010), when climate change was first addressed by the UN general assembly in 1988, the focus was on mitigation because climate change was perceived as a global problem requiring global collaboration. Adaptation, which at the time was understood to be local, was seen as inherently undesirable owing to concerns that some countries might choose as a matter of policy to invest in adaptation rather than mitigation due to its perceived lower costs, undermining the global coalition required to address climate change (see, Kjellen, 2006). Another concern, expressed most famously by Al Gore, was that adaptation represented a lazy and arrogant attitude to climate change; lazy because it avoids the hard work required to mitigate the problem and arrogant because it presumed that the problem could be solved on the back end i.e. after the fact (see, Pielke, 1999, p. 162). During the following decade however, climate change proponents came to accept what in hindsight should have been obvious from the beginning; that because GHGs are long-lived in the atmosphere and their effects cumulative, many projected impacts would proceed apace even in the improbable event that mitigation efforts completely halted emissions. Thus, it would appear that it was Mr. Gore who demonstrated at best unjustified optimism or worse, arrogance, in believing that the problem could be solved entirely or largely on the front end via curtailing emissions. In any event, following closely on this realization was another sobering revelation namely; that the impacts of climate change would be felt most strongly by the poor or developing nations of the world. Thus, in the emergence and acceptance of adaptation as a legitimate response to climate change, it was co-linked closely and strongly with developing countries.

2.1 Why are developing countries more vulnerable?

Climate change is expected to impact developing countries more severely because developing countries are more vulnerable. But what is vulnerability and why are developing countries more vulnerable than their developed counterparts? The IPCC (2007, p. 89) defines vulnerability as "the degree to which a system is susceptible to, and unable to cope with, adverse effects of climate change, including climate variability and extremes.

Vulnerability is a function of the character, magnitude, and rate of climate change and variation to which a system is exposed, its sensitivity, and its adaptive capacity." Thus, we can visualize this dependence graphically, using the analogy of the three legged stool (Fig. 1).

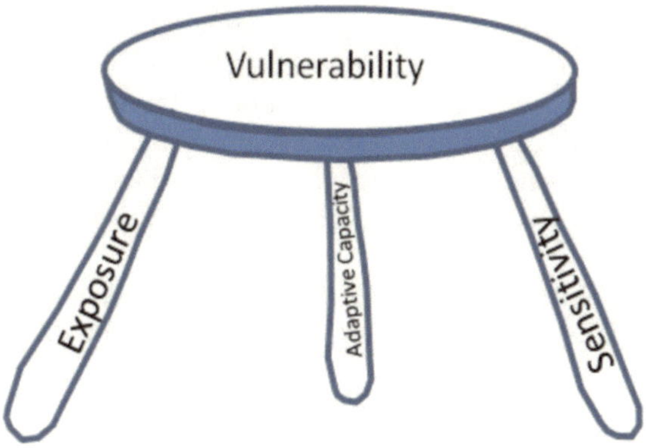

Fig. 1. Vulnerability as a function of exposure, sensitivity, and adaptive capacity.

The three-legged stool is a particularly useful analogy because inherent in the representation is the solution to managing vulnerability. The following paragraphs briefly review various dimensions of each leg of the stool.

2.2 Climate change exposure

Susceptibility to risks and hazards begins with exposure. That is, an entity (society or person) cannot be vulnerable to a risk to which they are not exposed. As an example, in order to be at risk of contracting tuberculosis, you must be exposed to the bacteria (*Mycobacterium tuberculosis*, MTB) or have been in close contact with a carrier of the bacteria. It is for this reason that public health officials recommend testing if you have travelled to TB endemic countries. But exposure to climate is a poorly defined term when it is defined at all. The IPCC glossary of climate change terms includes this rather unhelpful cyclical definition; "the degree to which a system is exposed to significant climatic variations" (IPCC, 2007, p. 373), which says exactly nothing. Public health officials are well aware that there is more to contracting a disease than mere exposure to it. A healthy, well nourished individual has a much higher likelihood of resisting a disease that would easily infect a weak individual with compromised immune system. There are in fact, three elements to exposure. The first is, of course, the climatic and environmental characteristics confronting the system. With respect to climate change this might include the expected degree of temperature rise. A region or country expected to experience a 4°C temperature rise has greater exposure than one where temperature is expected to rise by only 1°C. Coastal communities living in locations where sea level is expected to rise are exposed to the degree of expected sea level rise. The second element of exposure is the degree to which the system depends on the resource(s) in question. Egypt, for example, is nearly entirely dependent on the River Nile for its water resources. On the other hand, the Republic of Congo has abundant water resources and is relatively unconcerned about changes in the flow regimes of the Congo River. Third and finally, risks due to exposure are a function of system resilience or susceptibility. Resilience has a similar meaning to resistance; it refers to the amount of change a system can undergo without changing state. In climate change context, resilience also has been used to describe the ability of an individual, community or system to 'bounce back' and overcome adversity (Magistro, 2009).

SubSaharan Africa (SSA) agriculture offers textbook examples of exposure to climate change. As noted by Tarhule *et al.*, (2009), SSA depends more strongly and directly on rainfall than any other region on Earth. Approximately 65% of the labor force (FAO, 2006) and 95% of the land use (Rockström *et al.*, 2004) in the region are devoted to agriculture, and overwhelmingly to rain fed agriculture. In economic terms, agriculture contributes, on average, about 30% the Gross Domestic Product (GDP) for SSA countries (compared to 14% for developing nations elsewhere) and represents up to 55% of the total value of African exports (Sokona & Denton, 2001; UNFCCC, 2006). Seventy percent of the regions' labor is employed in agriculture but for the poor, that percentage is as high as 90%. Thus, Sub-Saharan Africa scores high on each of the elements that characterize exposure. Similar kinds of dynamics can be explicated for the region with respect to other climate change impacts including health, water, biodiversity, and coastal flooding. It is the culmination of all such domain exposures that make the region so highly vulnerable.

2.3 Climate change sensitivity

Now, suppose a system is exposed to a risk, then the degree to which the system is impacted or vulnerable depends on its sensitivity. Sensitivity is the degree to which a system is affected, either adversely or beneficially, directly or indirectly by, climate related stimuli (IPCC, 2007). If a system is highly sensitive or susceptible to a given risk, then it suffers the associated risk impacts commensurate with its degree of exposure. On the other hand, systems that are completely insensitive may experience zero impacts even though they may be completely exposed to the risk. Public health officials are often flummoxed by isolated cases of sex workers who never contract HIV (Human Immunodeficiency Virus) even after multiple exposures to the disease. These workers are immune or insensitive to the effects of the virus. To a degree, therefore, a system's sensitivity may completely counterbalance the effects of exposure.

Continuing with the illustration of African agriculture, we note that it is also highly sensitive to climate change, stemming from the fact that the vast majority of SSA agricultural production is near subsistence level and unable to produce substantial surplus even during years of good rainfall. Consequently, the region historically has experienced cycles of feast or famine that mimic the pattern of rainfall surplus and drought. In short, as the annual rainfall cycle goes so goes African agriculture and, along with it, the welfare and livelihoods of the populations that depend on agriculture for a living. Yet, climate change is expected to exacerbate existing patterns of climate. For Africa this change is expected to manifest primarily as increased temperatures, more variable rainfall characterized by more frequent occurrences of extreme events, such as droughts and floods, possible shortening of the growing season, and spatial shift in the growing zones of staple and economic crops, among other things.

2.4 Adaptive capacity

Adaptive capacity is the "ability of a system to adjust to climate change (including climate variability and extremes) to moderate potential damages, to take advantage of opportunities, or to cope with the consequences" (IPCC, 2007, p. 365). A rich and growing

literature on adaptive capacity (for an excellent and succinct review, refer to Smit & Wandel, 2006) has identified many of the determinants or drivers of the process (Figure 2). Some determinants are local while others reflect more general social political and economic systems. Local adaptive capacity "reflects such factors as managerial ability, access to financial, technological and information resources, infrastructure, the institutional environment in which adaptation occurs, political influence, kinship network, etc" (Smit & Wandel, 2006, p. 287). Determinants related to the general social political and economic systems include factors such as the availability of crop and flood insurance whether private or subsidized by the state, as well as the ability of impacted groups to influence regional or national policies related to their domain or sector.
In theory, even if a system or society is exposed to a risk to which it is sensitive such risk would have minimal adverse effect given perfect or complete adaptive capacity. The system simply adjusts itself to compensate for or accommodate the new circumstances in much the same way as a chameleon might change its colors to blend in with its environment. In reality, of course, there is no such thing as perfect adaptive capacity, just as completely insensitive systems are rare. But there are degrees of adaptive capacity not just in every society but also in relation to specific risks. A society or system might score very high on adaptive capacity in response to a specific risk but the same system may exhibit extremely low adaptive capacity when faced with a different kind of risk.

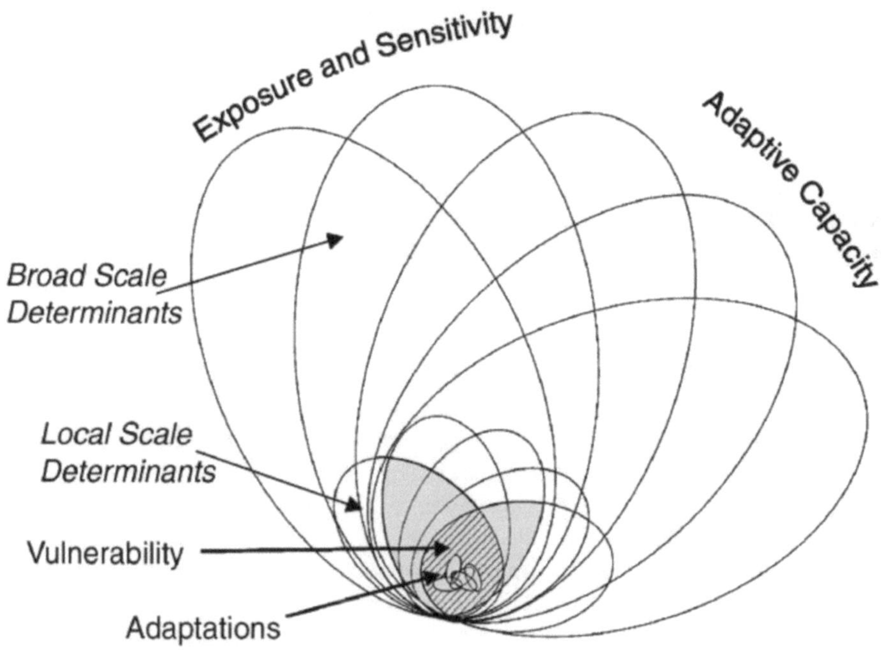

Fig. 2. A nested hierarchy and model of vulnerability (after Smit and Wandel, 2006). Reproduced with permission of the publisher.

Like any other region, Africa has attempted to adapt to its historical pattern of climate variability with some notable success stories. For example, Adger *et al.*, (2003) noted that agricultural communities in northern Nigeria continued to increase per capita agricultral production and stability during the period from 1970-2003 characterized by the longest and most intense drought in the history of the region as well as the confounding effects of population growth (see also, Nyong *et al.*, 2007). Elsewhere in Bangladash, local governement investments in shelters have helped reduce the mortality from cyclones (Adger *et al.*, 2003, p.186). Despite such successes, frequent and widespread droughts accompanied by massive famines, such as the Sahel droughts of 1970-73 and the East African drought of 1983-85, the Mozambican floods of 2000, the Sahel floods of 2007 (Paeth *et al.*, 2008) underscore the fact that adaptation to high intensity or high magnitude climate anomalies has generally been unsuccessful or inadequate. Similar examples of adaptation limitations or outright failure can be found in the areas of health, environmental and biodiversity resources, fish, livestock and coastal systems, among several others.

Although discussed separately for purposes of clarity, exposure, sensitivity and adaptive capacity are all strongly interconnected through the environmental, political, and socio-economic processes driving them (Smit & Wandel, 2006, p. 286). In other words, many of the same reasons for a systems high exposure underlie its sensitivity and low adaptive capacity. Furthermore, all three concepts are dynamic, context-specific, and scale-dependent (Smit & Wandel, 2006).

2.5 Conceptualizing the role and place of adaptation in development

Development experts have long argued that the elements of climate change vulnerability are essentially the same as the attributes that cause underdevelopment. Indeed, both developed and developing countries have various degrees of exposure and sensitivity to climate change in different domains. In fact, from a strictly environmental and climatic point of view, there are several areas in which climate change impacts are expected to be more severe in industrialized countries than in the developing countries. For example, many General Circulation Models (GCMs) predict higher proportionate temperature, precipitation, and ecosystem range changes in high latitudes (where most developed countries are) than in the low latitudes (where most developing countries are). But the greater focus and concern of both development and climate change scholars is on developing countries which will be least able to deal with the impacts of climate change. The logic that emerges from this consideration is the following: if development equals an enhanced ability to deal effectively with climate change, is accelerated development the solution to climate change adaptation in developing countries?

In fact, from the beginning of its recognition as a major global challenge, climate change has been linked to development. The 1987 Brundtland report *Our Common Future* noted explicitly that climate change constituted a major challenge facing development in poor countries (World Commission on Environment and Development, 1987). However, for the next decade and half following this declaration, the dominance of the mitigation agenda among the climate change research community prevented further exposition of the linkages and dynamics. Ayers & Dodman (2010) have posited that the absence of explicit mention of climate change adaptation in the Millennium Development Goals is because the document was drafted during this lull. In 2002, adaptation received a major boost when 10 leading development funding agencies released a report titled *Poverty and climate change: Reducing the vulnerability of the poor through adaptation* in which they argued that pro-poor development was key to the achievement of the Millenium Development Goals (MDGs) and successful adaptation.

Beginning with that opening, there are now three identifiable ways in which adaptation is perceived in the development context. These are stand alone adaptation, mainstreaming adaptation or adaptation plus development, and adaptation as development (Ayers & Dodman, 2010, p. 165).

Stand alone measures view adaptation strictly in terms of responses to climate change due to anthropogenic activities (UN, 1992). This approach stems from early definitions of climate change by the IPCC as due solely to human activities (i.e. anthropogenic global warming) rather than climate variability which includes also fluctuations due to natural earth system processes. In this sense, climate change is perceived to be what is additional to the baseline patterns of natural climatic variability. Consequently, the intervention strategies proposed also tend to be additional to baseline developments needs; those needs that would have existed within the community or society regardless of climate change are underemphasized or ignored completely. Thus, we see in this approach vestiges of early perceptions of adaptation as something local while climate change was considered a global phenomenon. Another characteristic of the stand alone approach is that it prioritizes technical and scientific intervention strategies such as dams, early warnings sytems and irrigation projects among others. It seeks to address vulnerability but not the underlying drivers of vulnerability, including questions of equity, access, and affordability.

This approach has been criticized for addressing the symptoms rather than the causes of climate change vulnerability. Critics point out for example, that an intervention strategy, such as a dam or reservoir while technically and conceptually sound may be of limited success against climate change vulnerability if vulnerable groups lack access to the dam for political, social or other reasons or if they cannot afford to pay for irrigation water for economic reasons. Another limitation of the approach is that it could actually act as a constaint to effective adaptation. Ayers & Dodman (2010) cite the example of Tuvalu, a poor island nation, where sea level rise and erosion are expected to exacerbate existing erosion problems. Clearly, both the current erosion problems and the anticipated future erosion need to be addressed. But by UNFCCC convention, the bottom part of the erosion control infrastructure needs to be built by the Tuvalun government as part of its normal development obligations. Tuvalu could then apply for funds through the National Adaptation Programs of Action (NAPA) designed to assist the least developed countries deal with climate change impacts to build the top part section i.e. that which would curtail erosion due to climate change. The problem is that Tuvalu is unable to build the bottom section because it does not have the money to do so with the consequence that the nation is unable to access funds for which it is eminently eligible.

The second approach attempts to mainstream adaptation into development strategies and policies or to 'climate-proof' development by ensuring that development interventions will be able to withstand the effects of climate change. This approach is favored by major funding agencies, including the World Bank, which in 2010 released its strategic framework on development and climate change (World Bank, 2010). The approach integrates climate change adaptation into social, institutional, and infrastructural development but has been criticized for depicting adaptation as something tacked on to development i.e. adaptation plus development (Ayers & Dodman, 2010). To a degree, this sweeping criticism is unfair because in some cases, these agencies deal with projects conceived or even partially implemented prior to the policy decisions to factor climate change effects. In such instances the agencies are simply retroactively climate proofing their development interventions.

The third and final approach views adaptation as synonymous with development (adaptation as development). Here, the goal is to address the general indicators of underdevelopment, including poverty, access to education, health services, finances, information, technology, improved living conditions etc, with a view to ensuring that the victims would be better able to deal with the effects of climate change. In a sense this approach implicitly attempts to reduce vulnerability in developing countries through fortifying them with the same attributes that make developed countries less vulnerable at least on the fundamental indicators. Diligent adherents of this position recognize that in some cases short-term or exigent development priorities might conflict with long term adaptation needs, leading potentially to maladaptation or conflict with local stakeholders. For example, confronted with an environmental or climatic hazard, many local and especially poorly educated stakeholders think in terms of immediate survival and coping rather than long-term adaptation. It is difficult to argue with this logic because the question of adaptation or long term impacts is moot if one is not around to deal with it. The problem arises if donor agencies, which have the resources and luxury of focusing on long-term adaptation insist on prioritizing adaptation over coping.

The view of adaptation as been synonymous with development begs further exploration. Literature search indicates that few studies have bothered to critically examine the question of why adaptation needs to be conjoined to development, as well as the follow up question of whether this linkage is real or contrived. There are good reasons for such exercise. For decades development scholars have promoted various strategies as panacea to underdevelopment, from technology transfer, to trade liberalization, to democracy, to debt forgiveness, and now... adaptation. To date, while there has been some progress, many developing countries particularly in SSA remain firmly in the grips of crushing poverty. What reason is there then, for believing that casting adaptation as development would be a more effective strategy; that it will have a better chance of bringing about sustained development where the previous policies have failed? Or might the approach, like its predecessars, fail to live up to its lofty expectations? In an attempt to address these questions, a good starting point is to contrast global efforts toward, and characterization - some might say the marketing- of climate change adaptation with the MDGs.

The MDGs are time-bound targets designed to provide concrete, numerical benchmarks for tackling extreme poverty in its many dimensions, through unified global efforts in education, health, environment, and economics (www.beta.undp.org/undp/en/home/mdgoverview.html). The United Nations Development Program (UNDP) which developed and oversees the project proclaims that the MDGs provide a framework for the entire international community to work together towards a common end, making sure that human development reaches everyone, everywhere (www.undp.org/mdg/basics.shtml). Thus the declared motive of the MDGs initiative is to ensure that *development* reaches everyone everywhere. Yet, today, it appears that it is adaptation to climate change, not the MDGs, that is being promoted as synonymous with development. In fact, to read the literature, the future survival of developing countries does not seem to be as depenendent on achieving the MDGs as it is on adapting effectively to climate change. Certainly, enhancing the development profile of the world's poorest countries is needed for its own sake so why does development need the imprimatur of adaptation?

One possible explanation is that climate change has broader impacts and threatens "all life on the planet" making it "the cause of anticipatory grief and felt loss" (Reser & Swim, 2011). In contrast, the MDGs have the most significance for developing countries even though a legitimate argument could be made that what ails the developing countries ultimately affects the developed nations. Nevertheless, what is true for social media and international discourse is also true for scholarly research, namely, issues that affect the developing countries exclusively or primarily tend to receive much less coverage. This fact may account for the seemingly greater attention to climate change than the MDGs with the latter forced to align itself with the former to share in the glow as it where.

Another way to interpret the seemingly low key international perception of the MDGs is because they address a set of concrete uncontroversial issues. Controversy breeds attention. Every 20 seconds, a child dies from water related disease; 2.5 billion people lack access to improved sanitation, including about 1.2 billion who have no facilities at all; worldwide, over 3 million people die each year due to malaria, and in 2009, over 2.6 million new HIV infections were reported, claiming 6,000 lives each day. These are real numbers, not projections, and relatively unpoliticized. The point is that the causes of poverty and underdevelopment as well as their effects are obvious for anyone to see. That poor people suffer disproportionately from HIV/AIDS, malaria, and child mortality or lack access to safe drinking water or access to education and health amenities or credit is not a subject for debate. There is no question that young girls especially in the developing world are being discriminated against in terms of access to education. Additionally, the funding mechanisms for MDGs though structurally complex are very specific and require little justification. Funding is provided for example to combat Tuberculosis, HIV/AIDS, maternal health and so on.

In contrast, climate change adaptation attempts to address something that for the most part has not yet happened. This is a considerably harder sell especially when viewed against the backdrop of other very real and immediate needs. Adaptation as development both legitimizes and elevates the urgency associated with climate change, earning it a place at the crowded table of international development needs. This view in no way contests the very real development challenges that climate change poses nor does it doubt that effective adaptation will greatly accelerate development of the affected regions. It simply acknowledges that whether serendipitous or by design, portraying climate change adaptation as development serves the agenda of funding agencies. But this is not a case of funding agencies manipulating the development agenda. The UNFCCC has been an active and willing partner in this construction. At the Seventh Conference of the Parties (CoP), in Kuala Lumpur, Malaysia, the IPCC through the UNFCCC established the Least Developed Countries Fund (LDCF) to assist developing countries to adapt to climate change. This mechanism was reinforced a year later at the 8th COP of Delhi, India. The LDCF has been a major factor in shaping the research agenda related to climate change adaptation.

One question which the development scholars promoting the adaptation as development mantra have yet to address in any great detail is whether there is a risk that by conjoining the two adaptation is likely to suffer the same maladies that have stunted development in general. For example, throughout the 1970s and 80s, the developed countries poured billions of dollars in overseas development assistance to third world countries. In some countries notably in SSA, the money failed woefully in achieving its stated goals because it was diverted, mismanaged, there was a lack of local or internal technical and personnel capacity, and lack of political will and commitment among numerous others. The lessons learned from such past experience should inform the design and structure of the current efforts but thus far, development scholars appear to be more focused on winning the intellectual and policy debate and have yet to turn their attention to fine detail and logistical issues.

3. Climate variability or climate change: the question of what to adopt to

Due to considerable uncertainty concerning climate change projections, no one quite knows exactly how climate change will manifest and therefore the best way for dealing with it. One approach for planning for climate change adaptation is to use the past (observed) climate variability as analogue, i.e. temporal analogue. Analogues may also be spatial, i.e. when one looks at societal responses to climate change and climate variability in another region with comparable climatic conditions as the area of interest (Adger *et al.*, 2003). Such analogues require consideration of patterns and episodes of observed extreme climate variability as well as the corresponding responses of social, agro-ecological and environmental systems. Not everyone agrees that the analogues approach is ideal or the most effective approach. The key objection is that the future climate may differ markedly from the past in critical respects, including patterns of seasonal distribution and statistical characteristics such as mean, variance, frequency, and distribution of extremes. Moreover, when researchers talk of past climate as analogue, they typically consider primarily the periods of extremes, such as droughts, floods, and heat waves, and attempt to map the societal response to them. While such information is useful, these events are intermittent and it is not always clear how the dynamics of the coping strategies would evolve on a sustained or protracted basis. Consider the Sahelian drought of 1970-73 or the East African drought of 1983-85; in what ways would the affected regions have adapted if those levels of climate variability had become permanent? Additionally, the onset of extreme events tends to occur over a short period of time, even for creeping type phenomena like drought, relative to climate change.

Despite such concerns, there are also strong reasons why analogues provide acceptable templates for adaptation and response to change. In some parts of the world, such as the African Sahel region, the range of observed climate variability exceeds the expected range of future variation due to climate change.

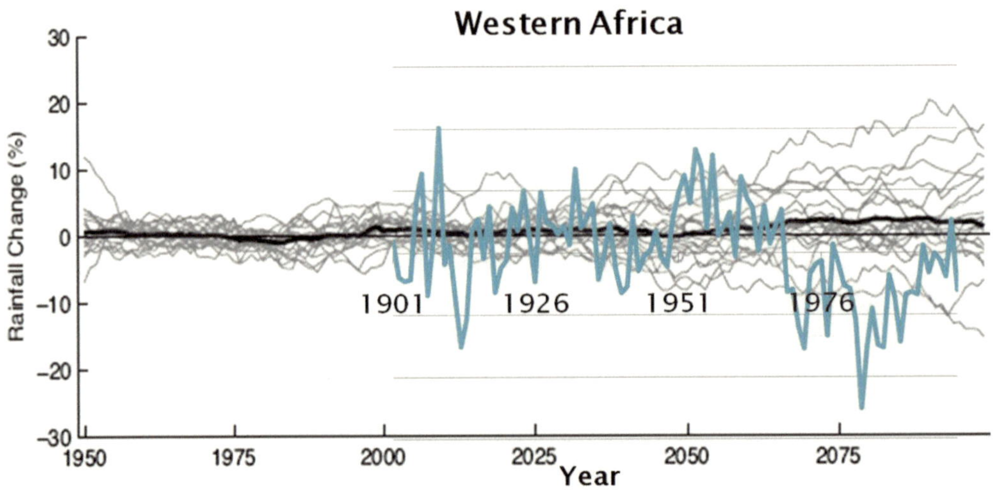

Fig. 3. Plot of observed climate variability (turquoise blue) during the 20th Century superimposed on GCM projections for the 21st Century showing percent rainfall changes in West Africa. (Courtesy, Casey Brown, University of Massachusetts).

In Fig. 3, the researchers superimposed the observed pattern of rainfall changes in West Africa during the last century over an ensemble plot of GCM projections of rainfall changes for the present century. It is immediately obvious that the range of observed historical variability far exceeds the total range of the ensemble projections. In numeric terms, the Sahel region of West Africa experienced repeated and sustained periods of variability that ranged from about 25 to 40% depending on location (Hulme, 2001; Lebel et al., 2003). In contrast, projected rainfall changes over West Africa as a result of climate change are expected to be in the range of ±10%. By 2050, projected decline in runoff will be on the order of 5% compared to up to 60% during some periods of the last century. Also, the inability of GCMs to produce variability is well known. Even so, the significance of the example illustrated here goes well beyond either of those two caveats.

It is important to note the above situation for the Sahel is by no means unique. Observed rainfall variability has been shown to be greater than changes suggested by climate models for the next 50-100 years elsewhere in Africa, including eastern Africa and the Blue Nile Basin (Hulme, 1998; Adger *et al.*, 2003). Rainfall variability in northeast Brazil also is of the order suggested by future climate change (see, Marengo, 2004).

The key point is that for some parts of the world, the magnitude of historical climate, specifically rainfall variability that vulnerable groups have had to cope with and adapt to exceed the expected magnitude of climate change. Therefore, if vulnerable groups learn to cope with climate variability, they will be climate change ready.

The above suggests a fundamental flaw in the UNFCCC approach and funding mechanism, which seeks to focus adaptation efforts on climate changes above baseline conditions. Clearly, helping vulnerable groups to deal effectively with current climate variability, which means addressing the determinants of vulnerability would appear to be the superior approach. These examples emphasize the point that adaptation is very context specific, a fact that is frequently acknowledged but just as frequently glossed over.

4. Part 3: Implementing adaptation to climate change and climate variability

Researchers recognize several forms or levels of adaptation (Smit & Wandel, 2006). In terms of timing or when it is implemented, adaptation may be characterized as anticipatory i.e. carried out in advance of expected stimulus, or reactive i.e. in response to stimuli that has already occurred. It may be planned (formal) or spontaneous (informal), and it may be carried out at the level of the individual (private or autonomous) or an institution (public). This type of nomenclature has helped to impose a level of intellectual clarity regarding the different classes and types of adaptation responses. It also allows researchers to make informed guesses about the kinds of adaptation that vulnerable groups are most likely to implement under various contexts and scenarios. A review of the literature and consideration of relevant dynamics suggest the following.

4.1 Most adaptation in developing countries is likely to be autonomous and informal

In general, it can be expected that adaptation carried out by large entities such as governments and international development agencies are more likely to be planned and anticipatory. The reason is simple; these entities have access to the information, knowledge base, resources, institutional culture and even requirement for long term planning that is the hallmark of anticipatory adaptation. For example, the World Bank's strategy document on climate change adaptation (World Bank, 2010) requires all major World Bank funded projects to undertake climate risk assessments covering the life term of the project. UNFCCC requires NAPAs to factor in the risk of climate change in all of their planning documents. Yet, a major characteristic of developing countries is that the formal organized or planned sectors are weak and cover only relatively small portions of the overall population and activities. And there is unlikely to be sufficient development aid or investment to implement planned adaptation in developing countries. The UNFCCC acknowledged as much when it stated, "it is clear that current funding is not enough to support adaptation needs. Recent studies...showed that an incremental level of annual investment and financial flows of about $50 billion is needed for adaptation in 2030. In the context of any discussion on future international cooperation on climate change, future financial resources need to be sufficient, predictable and sustainable in order to facilitate adaptation to adverse impacts of climate change by developing countries" (UNFCCC, 2007, p. 52). Therefore, it can be expected that public, formal, and planned climate change adaptation interventions will represent only a small proportion of the overall climate change adaptation activities needed in these countries. The balance of the interventions necessarily will have to be made up through individual efforts. But a poor African farmer, for example, typically does not think about climate in 2030; he/she is much more likely to be concerned with the next year or two and 5 years would be considered a long planning horizon (see also, Salick & Byg, 2007).

The above consideration suggests that climate change adaptation planning in developing countries must recognize the reality that significant amount of adaptation will be reactive, informal, and private. It bears to note that there is nothing wrong with these modes of adaptation. We have already made the point that adaptation needs to be context specific and dependent on the resources available to the impacted domain. Therefore, for poor, small scale farmers, the actions that they take in response to climate variability and change are the actions that they can afford or which are most relevant to their needs. Moreover, the lines between anticipatory adaptation and reactive adaptation are not always rigid or even clear. For example, Reser and Swim (2011, p.284), cite the case of an individual who rebuilds his home after a storm. This individual may decide to build the home stronger and away from likely storm or flood damage. In this example, the act of rebuilding the home is clearly reactionary. Yet the decisions about how to make the home less exposed or susceptible to future storm damage is clearly anticipatory.

A number of actionable policy guidelines emanate logically from the foregoing discussion. First, recognizing that planned adaptation will not reach all groups and sectors that need it, there is a need to ensure that those groups which will be forced to act on their own have the right information on which to base their actions. Without such information, impacted groups will have no option but to rely on past experience, which may not be a good predictor of future outcomes. Indeed, a major risk to adaptation or cause of maladaptation, is when people act on the basis of faulty information and assumptions. Tarhule (2005), cites the example of responses to drought in the Republic of Niger. Following three decades of declining rainfall and diminishing water resources, many farmers relocated their homes and activities within or close to flood plains in order to be closer to water. On the face of it, such decision appears reasonable but once wetter conditions returned, these activities were confronted with greater flood risk directly as a result of their drought response strategies. It is reasonable to expect that climate change may entail many such incidences in the future because past experience will no longer be an adequate and sufficient predictor of the future. There is also the risk of disruptions to expected trends caused by possible non-linearities in climatic patterns. The key message here is that the governments and development agencies working in developing countries need to provide information that guides vulnerable groups about sustainable and effective adaptation practices as well as maladaptation risks.

Moreover, it is important to ensure that vulnerable groups can in fact act upon that information. Generic information about what individuals and at-risk communities should do is useless if those communities cannot put such actions into practice. Buchanan-Smith *et al.* (1994) in their provocatively titled article, "let them eat information" provide an excellent discussion of the great gulf that exists between information and people's ability to put that information to beneficial use. Therefore, adaptation planning and policies need to be smart in identifying the kinds of interventions that are needed or likely to be most effective in specific context. Research in livelihoods analysis and consideration of the status of adaptive capacity within communities provides some guidance. For example, Fabricius *et al.*, (2007), utilized a livelihoods analysis framework to define three broad types of communities in terms of their adaptive capacity. These include (see Magistro, 2009):

i. *Powerless spectator* communities with institutionalized histories of powerlessness and disadvantage have a low adaptive capacity and weak capacity to govern, do not have financial or technological options, and lack natural resources, skills, institutions, and networks.
ii. *Coping actor* communities have the capacity to adapt, but are not managing social-ecological systems. They lack the capacity for governance because of lack of leadership, of vision, and of motivation, and their responses are typically short-term.
iii. *Adaptive manager* communities have both adaptive capacity and governance capacity to sustain and internalize this adaptation. They invest in the long-term management of ecosystem services. Such communities are not only aware of the threats, but also take appropriate action for long-term sustainability.

Knowing the status or adaptive capacity of each comunity or domain can help in the development of the adaptation strategies most suitable to their situation and their abilities. For adaptive manager communities, all that might be necessary is a little bit of information or assistance and they will be on their way. For coping communities, greater emphasis may need to be placed on building their capacity to respond as a prerequisite to engaging in adaptation activities. Powerless spectator communities also need a healthy dose of capacity building but for many of these communities, there may be no alternative to direct intervention to help stave off expected suffering and economic loss.

Thus, vulnerability mapping needs to take into account not only expected climate change impacts but also community mapping and designation in terms of adaptive capacity. While conceptually sound, developing the metrics to capture sufficiently robust and dependable adaptation characteristics to permit such mapping would be challenging. But this is the kind of research that the hazard and, increasingly, development communities, have proved themselves most capable of.

4.2 Individuals and vulnerable communities and indeed many developing countries will have little motivation to act on climate change unless they perceive real benefits

That is, vulnerable groups are unlikely to implement some adaptation strategies regardless of the logic or soundness of such strategies unless those strategies address an immediate need or a recognizable future threat. The reason has to do with the razor thin margin of resources and efforts available to many vulnerable groups. Faced with deploying resources to address some immediate need and some alleged future threat, most people will, and rightly so, prioritize the more immediate threat. This point underscores the fact that adaptation will be very context specific. Consequently, policy makers and development planners who seek to mainstream adaptation into current project must be prepared to show clear and direct linkages.

4.3 There may be no grand theory of adaptation. A cursory examination of the rather copious body of work on adaptation to climate change and climate variability shows that there is very little specificity in the recommendations that most studies offer

Even the UNFCCC adaptation (see, UNFCCC, 2007) strategies appear short on specificity despite being very detailed on problem description and analysis. On the other hand, the database on local coping strategies (http://maindb.unfccc.int/public/adaptation/) is potentially quite useful, not because it is more specific but because it simply documents coping strategies elsewhere, providing users insights into how other communities are learning to cope with climate change and climate variability.

The reason for the lack of specificity has to do with the previously discussed characteristics of the adaptation process; it is dynamic, context specific, and scale-dependent. Perhaps, in the end, one of the most effective adaptation strategies will be to teach vulnerable communities the *process* of doing adaptation. For example, at the most general scale we know that climate change and climate variability risks are caused by exposure, sensitivity, and adaptive capacity. Therefore, the solution is to reduce exposure, reduce sensitivity, and increase adaptive capacity. Beginning with this framework and working to progressing finer levels of detail and spatial resolution one might attempt to figure out for a given location who or what is exposed to climate variability and to what degree. Resolving those issues might prompt questions about why such exposure exists and finally how it might be mitigated. One could then follow the same *process* of reasoning with respect to how to reduce sensitivity or how to increase adaptive capacity. A focus on the process and on information dissemination puts the stakeholders in the drivers' seat rather than treating them as passive and helpless victims, but see below.

4.4 There is a need for more studies of the psychology of climate change adaptation and motivational behavior

Nearly all climate adaptation strategies implicitly assume that "the reduction of appraised threats motivates individuals to initiate coping responses" (Reser & Swim, 2011, p. 283). This assumption is not always rigorously tested or even investigated, especially in developing countries. Most people are aware of the health dangers that smoking poses but that does not stop them from indulging the habit. Other people also voluntarily engage in risky sexual behaviors and drug habits, fully knowing the risks involved. In Africa, efforts to combat issues like high population growths rates, land degradation, and female education often have run up against social-cultural norms and perceptions that are very difficult to overcome. What motivates people to decide to act or not act or choose the actions that they do in the face of clear and present danger? What are the emotional and cognitive heuristics involved and how might lessons learned from combating other threats like anti-smoking campaigns inform the climate change and climate adaptation efforts. These types of questions have not been sufficiently posed and investigated.

4.5 Adaptation has limits

Some sectors and activities are inherently structurally defective to the point where their adaptation options are limited. Other sectors and activities are doomed by timing and circumstances and therefore unfit for adaptation. Pastoral nomadism is a good example of a system that is doomed by external circumstances beyond redemption. Each year, nomads find their traditional migratory routes blocked by land use changes as well as population and economic growth. There is no help for this sector other than total abandonment, even without the confounding effects of climate change. The best analogue for pastoral nomadism is the open range concept, which was prevalent in the United States during the 2nd half of the 19th Century. Open range referred to areas of public domain that were used as common pasturelands for cattle grazing. Fencing was prohibited on these lands, allowing free movement of cattle. The open range concept became unsustainable as the USA became more and more settled and economic growth necessitated changes to the land tenure system. Such is the case in parts of Africa today, where conflicts between nomadic livestock herders and sedentary farmers have intensified in recent years (Tarhule, 2002).

In a similar vain, Adger *et al.* (2003, p.189) noted,"the limits to many adaptation options are already apparent in areas such as population movement and migration, in the ability to bring new agricultural land under irrigation when rainfall is threatened, or to bring about large-scale infrastructural changes to minimize the impacts of sea-level rise on coastal areas." These examples make the point that recognition and even willingness to adapt may not be synonymous with the ability to do so. In some instances the most appropriate response to climate change and climate variability may be wholesale abandonment of lifestyles and sectors although some people may argue that such steps, too, represent merely an extreme form of adaptation.

5. Summary and conclusions

Climate variability and change will constitue, arguably, the defining phenomenon of this century. Societal response to climate change comprises two sets of activities that lie on the opposite ends of a continuum. The first is mitigation, which attempts to curtail the entry of additional GHGs into the atmosphere as a result of human acitivities. The second is adaptation, which attempts to increase the capacity of human and agro-ecological systems for dealing with the changes caused by climate change. Almost by tacit agreement, it is understood that developed countries will focus on climate change mitigation while developing nations will focus on climate change adaptation. This arrangement makes sense from a variety of perspectives. It is the developed nations which have contributed the most to the current GHG concentrations in the atmosphere and which also have the technological and economic resources for tackling the problem of emissions reduction. Developing nations, on the other hand, have generally contributed less to GHG emissions (although some of the larger emerging nations like China, India, and Brazil recently have became major GHG emitters). Most developing nations also contain significant populations and domains that are especially vulnerable to the impact of climate change, hence it makes sense that they should focus on ways of reducing that vulnerability.

Drawing from a variety of sources, including ecology hazards and development studies, a rich body of literature has emerged providing intellectual clarity on the philosophy and approach to climate change adaptation as well as the nuances and meanings of associated concepts. While much progress has been made on the academic front, the process of translating theoretical constructs into implementable ideas has lagged. This chapter undertook a critical review of the situation, mainly from an African perspective, and identified a number of salient factors necessary for bridging the gap between adaptation theory and practice, in other words, how to move beyond rhetorics.

The literature suggests that a considerable amount of research efforts has been devoted to explaining the role of adaptation in development. The key questions may be summarized as follows. Should adaptation focus on the climate change effects over and above baseline conditions or should it target the factors that make people vulnerable to climate change? When does adaptation equal development? What kinds of development qualify as adaptation? While stimulating intellectually and perhaps from the view point of how one funds adaptation, the ultimate value of these lines of analysis may be most pertinent with respect to planned or institutional adaptation activities. For most developing countries however, adaptation is likely to be piecemeal, autonomous, and reactive simply because there will not be sufficient money or technical capacity or personnel resources to reach a majority of the people who need help adapting to climate change. This realization suggests that adaptation efforts should focus on giving peope the information they need to help themselves.

Development experts should be careful not to oversell the potential of climate change adaptation in bringing about development. Many such promises have been made in the past, including the Green Revolution, technology transfer, and debt forgiveness, and democracy but all ultimately fell short of the promised dividends. Instead, it may be more beneficial to focus on the lessons learned from those past experiences and failures, with a view to ensuring that adaptation lives up to its billing.

A number of assumptions regarding adaptation have not yet been sufficiently tested. The assumption that people are motivated to act once they have information may need to be tested. Similarly, not enough studies have investigated the psychology of climate change adaptation or even how social-cultural beliefs and practices may hinder (or promote) climate change adaptation. This is important considering the degree of inertia faced by other major social transformation initiatives, like anti-smoking campaigns or birth control in developing countries.

For many developing countries especially those in Africa, it would appear that climate variability and not climate change is the more serious threat. In some locations, the range of observed climate variability during the past century exceeds the expected magnitude of climate change. This suggests that if the impacted regions can cope with climate variability they will be climate change ready.

Finaly, developing countries need to assume greater ownership over the climate change adaptation process to ensure both its sustainability and management according to national development priorities.

6. References

Adger, W. N., Huq, S., Brown, K., Conway, D., & Hulme, M. (2003). Adaptation to climate change in the developing world. *Progress in Development Studies* , 179-195.

Ayers, J., & Dodman, D. (2010). Climate Change Adaptation and Development I: the state of the debate. *Progress in Development Studies* , 161-168.

Buchanan-Smith, M., Davies, S., & Petty, C. (1994). Food Security: Let them eat information. *Intitute of Development Studies Bulletin* , 1-18.

Fabricius, C., Folke, C., Cundill, G., & Schultz, L. (2007). Powerless spectators, coping actors, and adaptive co-managers: a synthesis of the role of communities in ecosystem management. *Ecology and Society*.

Food and Agriculture Organization of the United Nations (FAO). 2006. *FAOSTAT Online Statistical Service.* Rome: FAO. Available online at: HYPERLINK "http://faostat.fao.org" \t "_blank" http://faostat.fao.org .

Hulme, M. (2001). Climatic Perspectives on Sahelian desiccation: 1973-1998. *Global Environmental Change* , 19-29.

Hulme, M. (1998). The sensitivity of Sahel rainfall on global warming: implications for scenario analysis of future climate chage impact. In E. Servat, D. Hughes, J. Fritsch, & M. Hulme, *Water Resources Variability in Africa During the 20th Century* (pp. 429-436). Wallingford: IAHS Publication.

IPCC. (2007). *Climate Change 2007: Synthesis Report. Contribution of Working Groups I,II, and III to the Fourth Assessment Report of the Intergovernmental Panel on Climate Change.* Geneva, Switzerland: IPCC.

Kjellen, B. (2006). Forward. *In* Adger, W.N. , Paavola, J. , Huq, S. and Mace, J. , *editors, Fairness in adaptation to climate change.* MIT Press .

Lebel, L., Redelsperger, J.-L., & Thorncroft, C. (2003). African Monsoon Interdisciplinary Analysis (AMMA): An international research project and field campaign. *CLIVAR Exchanges* , 52-54.

Magistro, J. (2009). *Coping with climate risks: An Africa review.* Tucson: Bureau of Applied Research in Antrhopology.

Marengo, J. (2004). Interdecal variability and trends of rainfall across the Anmazon Basin. *Theoretical and Applied Climatology* , 79-96.

Nyong, A., Adesina, A., & Elasha, B. O. (2007). The value of indigeneous knowledge in climate change mitigation and adaptation strategies in the African Sahel. *Mitigation and adaptation strategies for global change* , 787-797.

Paeth, H., Fink, A. H., Pohle, S., Keis, F., Machel, H., & Samimi, C. (2008). The 2007 flood in sub-Saharan Africa: spatio-temporal characterisitcs and potential causes. *International Journal of Climatology, DOI:10.1002/joc.2199* .

Pielke, R. (1999). Nine fallacies of floods. *Climatic Change* , 413-438.

Reser, J. P., & Swim, J. K. (2011). Adapting and coping with the threat and impacts of climate change. *American Psychologist* , 277-289.

Rockström, J., C. Folke, L. Gordon, N. Habitu, G. Jewitt, F. Penning de Vries, F. Rwehumbiza, S. Sally, H. Savenije, and R. Schulze (2004): A watershed approach to upgrade rainfed agriculture in water scarse regions through water system innovations: an integrated research initiative on water for food and rural livelihoods in balance with ecosystem functions. Physics and Chemistry of the Earth, 29, 1109-1118.

Salick, J., & Byg, A. (2007). *Indigenous Peoples and Climate Change.* Oxford: Tyndall Center for Climate Change Research.

Smit, B., & Wandel, J. (2006). Adaptation, adaptive capacity and vulnerability. *Global Environmental Change* , 282-292.

Sokona, Y., & Denton, F. (2001). Climate change impacts: can Africa cope with the challenges? *Climate Policy* , 117-123.

Tarhule, A., Z. Saley-Bana, and P.J Lamb (2009). Rainwatch: A Prototype GIS for Rainfall Monitoring in West Africa. *Bulletin of the American Meteorological Society,* 1607-1614.

Tarhule, A. (2005). Damaging rainfall and floods: the other Sahel Disasters. *Climate Change* , 355-377.

Tarhule, A. (2002). Environment and conflict in West Africa. In M. Manwaring, *Environmental security and global stability: problems and responses* (pp. 215-237). Lexinglton Books.

United Nations. (1992). Report on United Nations Conference on Environment and Development. http://www.un.org/documents/ga/conf151/aconf15126-1annex.htm.

UNFCCC. (2007). *Climate change: Impacts, vulnerabilities, and adaptation in developing countries.* United Nations Framework Convention on Climate Change.

UNFCCC. (2006). Background paper on impacts, vulnerability, and adaptation to climate change in Africa. *Workshop on Adaptation Implementation of Decision 1/CP.10 of the UNFCCC Convention.* Accra, Ghana, 21-23 September: UNFCCC.

World Bank. (2010). *Development and climate change: Stepping up support to developing countries.* Washington DC: World Bank Group.

World Commission on Environment and Development. (1987). Our common future: Report of the World commission on Environment and Development. Annex to General Assembly Document A/42/47, *Development and International Cooperation: Environment,* 2.

Yohe, G., Malone, E., Brenkert, A., Schlesinger, M., Meij, H., & Xing, X. (2006). Global distributions ofvulnerability to climate change . *The Integrated Assessment Journal* , 35-44.

9

Adapting Agriculture to Climate Variability and Change: Capacity Building Through Technological Innovation

Netra B. Chhetri
School of Geographical Sciences & Urban Planning
Consortium for Science, Policy, & Outcomes,
Arizona State University, Tempe, AZ,
USA

1. Introduction

Agricultural adaptation is an ongoing and dynamic process through which agrarian societies adapt to changing socioeconomic, technological, and resource regimes. Historically farmers have learned to thrive in a wide range of climatic conditions, ranging from extreme cold to hot temperature and from very dry to humid climate (Easterling et al., 2004). Variations in climatic resources across space and time have also acted as source of technological and institutional innovations (Hayami and Ruttan, 1985; Chhetri and Easterling, 2010). Human managed systems are continually responding to climatic conditions (Jodha, 1978), exhibiting their inherent potential for adaptation to future climate change. Examples of agricultural adaptation to climate also include translocation of crop across thermal gradients (Easterling, 1996; Gitay et al., 2001), substitution of new crops for old ones (Smithers and Blay-Palmer, 2001) and innovation of technologies to alleviate climatic constraints (Chhetri and Easterling, 2010).

Agricultural adaptation to emerging threat brought about by changing climate is entirely a different matter. Climate variability and change, especially the extremes, may create a new level of risks outside the realm of past experiences such as more frequent and extensive drought, abrupt break in seasonal rainfall, and extreme heat (Adger et al., 2007). It appears important, therefore, to seek to enhance the capacity of farmers to respond and adapt to such impacts of climate change. Agricultural adaptation, however, can be undertaken by range of stakeholders including farmers, public institutions, communities, civil society (NGOs) and private sectors. A sustained effort to adaptation in agriculture therefore demands an active engagement of various stakeholders so that location-specific technology is innovated to adapt to climate change.

While studies consistently show that climate change may not imperil the ability of world's agriculture to maintain food security in the short term (Gitay et al., 2001, Fischer et al., 2005, Easterling et al., 2007), it may challenge farmers to adapt in regions where it could be stressful and where new climate-induced opportunities remain unanswered. Agricultural adaptation, as a response to climate change, is contingent upon various factors including agro-ecological thresholds; values and cultural norms; agricultural policies and markets;

and the availability of technological choices for farmers (Reilly et al, 1996; Heyd and Brooks, 2009; Chhetri and Easterling, 2010).

Technology can potentially play an important role in adaptation in agriculture, yet much of what is known about the process of technological innovation in agriculture has yet to be captured in discussions of climate change adaptation. The time may now be opportune for a much broader debate on how technology might help adapt. This paper begins by discussing the role of technology in modern agriculture to illustrate the means through which society may respond to minimize the impacts of climate change in agriculture. This is followed by discussion of the values of climatic sensitive technological innovation in agricultural adaptation in the future. More specifically this section argues need for a close coupling of climate and technology so that location-specific technologies are delivered to and adopted by farmers. Fourth section illustrates the importance of understanding adaptations through analogy so that effectiveness of the adaptations practices to past climatic extremes across space and time can be examined. While the examples of farmers' response to extreme events of the past may not be directly applicable to a future adaptation, the insight may be gained about how the adaptation process may unfold in the future. In section five the premise of the hypothesis of induced innovation is discussed as it suggests an important pathway for the interaction of climate and technology and for the study of the agricultural adaptation to climate variability and change. The concluding section of this paper emphasizes the significance of technological innovation in agricultural adaptation in the future.

2. Role of technology in modern agriculture

While climate change poses a fundamental challenge often outside the range of current and past experiences, it is reasonable to expect that farmers and their supporting institutions would respond to new crop growing environments brought about by climate change the same way they have responded to climatic limitations of the past (Chhetri and Easterling, 2010). Insights for agricultural adaptation that confront us today may well be found in the experience of handling climatic challenges across various regions. Yet the science of adaptation has not progressed to the point where we have fair understanding of factors that enables smooth adaptation to new climate of the future. Much of what is known about the process of technological innovation has yet to be captured in discussions of agricultural adaptation to climate change. It is true that technological innovation in agriculture does not evolve with respect to climatic conditions alone; non-climatic forces, such as market and other social factors, clearly play a significant role. Yet research efforts in understanding the processes of technological change driven by climatic factors are pivotal to make any assertion about likely adaptation of agriculture to climate change.

Innovation, a process through which new (or improved) technologies are developed and brought into widespread use, have been, and will continue to be, central to adapting to climate change. For example, not only new and improved agricultural technology is crucially needed to adapt to changing climatic means and associated variability, it is also a solution space for addressing the food security challenge of ever growing population of the world. Additionally, new and carefully designed technologies may enable society to increase their robustness to tackle the emerging challenges emanating from climate and other ongoing changes. Innovation of technologies, however, is a nonlinear process. It is the product of constant interaction and feedback between social space (where individual interact) and organizational space mediated by infrastructure wherein individual and institutions operate. Therefore technological innovation involves the engagement of institutions and individuals in this space that cannot be trivialized. Depending on the nature and type, the process of technological innovation involved different institutional arrangement. In agricultural systems, for example, technological innovation has resulted in a more complex and sophisticated human-environment relationship in rice production. According to Chhetri and Easterling (2010), research establishment Nepal has used climate as one of the drivers of research and development of technologies in agriculture whereby multiple stakeholders, including farmers and NGOs, have worked together to develop technologies that consider local needs and climatic conditions.

A study by Smithers and Blay-Palmer (2001) also shows the role of farmers and their supporting institutions in alleviating the constraints of climate to soybean in Canada. A fundamental climatic constraint to soybean cultivation in Ontario, Canada was the prevalence of cold night temperature during its flowering period, confining soybean cultivation to the extreme southwestern portion of the province. A key innovation to address this constraint was the introduction of cold-tolerant genetic material, *Fiskeby63,* from Sweden that led to the development of *Maple Arrow* cultivar that eventually played a vital role in the eastward spread of soybean crop. Between 1970 and 1997, the total acreage planted to soybean increased by over 500 percent, with the expansion being attributed to a series of technological innovations made in response to the climatic condition of Ontario (Smithers and Blay-Palmer, 2001). According to Smithers and Blay-Palmer (2001), technological innovations were not only confined to development of cultivars but also to a range of agronomic activities including modification of planting time and crop rotation interrupted pest cycle enhancing the cultivation of soybean.

The development of cowpea cultivars in the African Sahel illustrates the example of technological substitution in response to existing variability in climatic resources. To escape the effects of drought, scientists in the African Sahel developed early maturing cowpea cultivars with different phenological characteristics. To avoid the effects of late season drought, they developed two cowpea varieties (*Ein El Gazal* and *Melakh*) that mature between 55 - 64 days after planting (Elawad and Hall, 2002). Similarly, to avoid mid-season drought, scientists also developed a cowpea variety (*Mouride*) that matures between 70-75 days after planting (Cisse et al, 1997). Unlike *Ein El Gazal* and *Melakh,* that begin flowering between 30-35 days from sowing and have synchronous flowering characteristics, the *Mouride* variety starts flowering in about 38 days after planting and spreads out over an extended period of time, thereby escaping the midseason drought. Additionally, to enhance the chances of significant grain production, agriculturists in this region developed an innovative cropping technique where both types of cowpea (short and medium maturing) are planted together so that the climatic input is optimized (Hall, 2004). Infrastructural innovation like the development of efficient irrigation infrastructure has been very effective in stimulating growth in agricultural production (Pinstrup-Andersen, 1982). In South and Southeast Asia, outcomes of the Green Revolution, characterized by development and diffusion of high yielding varieties (HYVs) of food crops, made possible by the availability of irrigation water, and early maturing HYVs facilitated by wider distribution of chemical fertilizers to maintain productivity, are considered to be positive developments, despite criticisms regarding distributional inequities and environmental concerns (Evenson and Gollin, 2000).

3. Sensitivity of technology to climate

Sensitivity of technological innovation to climate can serve as a potent means to adapt to progressively changing climatic mean and variability. The crux of the problem is whether or not climate variability and change serves as a driver of appropriate technological change. While ongoing research on climate change and agriculture is deepening the level of understanding, much of what is known about the process of technological innovation in agriculture has yet to be captured in discussions of climate change adaptation (Chhetri and Easterling, 2010). While non-climatic forces, such as market and other social factors, clearly play a significant role (Brush and Turner, 1987; Liverman, 1990), research efforts in understanding the processes of technological change driven by climatic factors are pivotal to making any assertion about likely adaptation of agriculture to climate change (Rosenberg, 1992). For example, research shows that the patterns of technological innovation in agriculture have generally served to reduce the dependence on the scarcest resources in each country where high rates of productivity growth have been observed during the 20th Century (Hayami and Ruttan, 1985). This experience suggests that if climate change causes a change in resource scarcities, public institutions do have some capacity to respond to those evolving scarcities. It is important to note here that for the assessment and analysis of climate adaptation, much of the agricultural technology that was developed in response to resource scarcity has been largely ignored. It is crucial to understand how climate and technology are interacting in agriculture so that a reasonable prognosis can be made about the possible adaptation to impending challenges of climate change.

Climate-technology interaction occurs when location-specific technologies are delivered to and adopted by farmers to address climatic limitations or opportunities in crop growth and development. This interaction is vital because greater crop yield is desired by all farmers regardless of the range of climatic conditions in which they farm. For example, an optimal cultivation of rice is possible only when climatic conditions are favorable that invariably means an inadequate supply of climatic resources (e.g. rainfall) adversely affects productivity. It is important therefore to understand how climatic limitations provide incentives for farmers to urge the institutions that support them to invest in research and development of location-specific technologies that substitute for climatic scarcity. The implication of this assertion is that regions that do not have location-specific technologies and/or infrastructure in place would not be able to cope and adapt to the changing future climate.

Translating this argument, as presented in the conceptual model (Figure 1), the induced innovation hypothesis suggests an important pathway for the interaction of climate and technology and for the study of the agricultural adaptation to climate change. The next section this paper will take the discussion further to include the role of technological change in climatic adaptation through the conceptual framework of the hypothesis of induced innovation. The justification for this approach is built on the premise that in order to know how well society might be prepared to adapt to changing climatic conditions, it is important to understand how well society has instituted technologies to address uncertainties emanating from climatic and other livelihoods stressors in the past, and consequently to minimize the risks associated with it. In addition, exploring the ways in which climatic uncertainties (unknown) induce or hinder technological innovations as means to respond to climatic constraints can have important implications for the analysis of agricultural adaptation in the future.

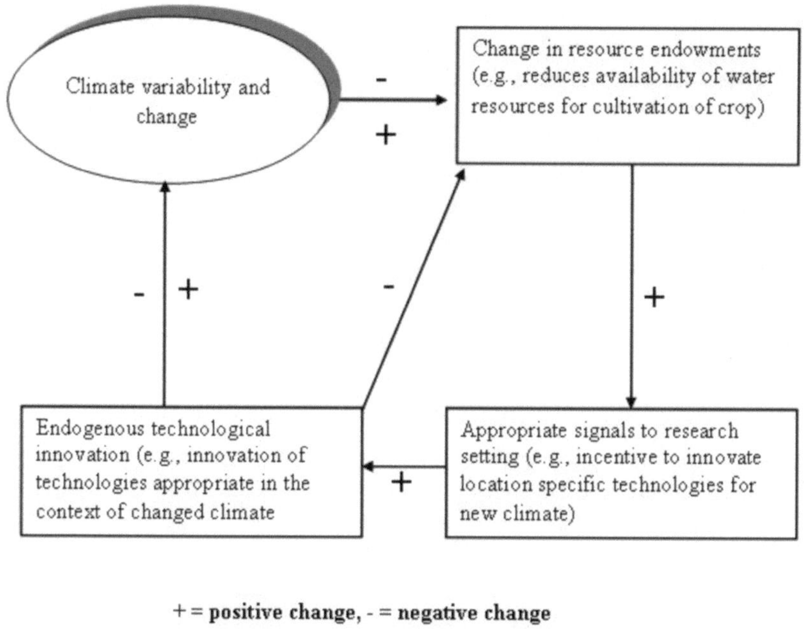

Fig. 1. Induced technological and institutional innovations: a conceptual perspective.

4. Understanding adaptation by analogy

One of the challenges in understanding the impacts of climate change on agriculture is gaining insights of how farmers and their supporting institutions interact to adapt to changing climatic conditions. Understanding adaptations through analogy may help evaluate the effectiveness of potential adaptation strategies by comparing observed adaptations to past climate extremes across space and time. Usually events that have had a similar effect in the recent past are compared to the likely impact of future events associated with climate change, assuming that lessons can be learned from such past experience and then applied to future situations. Although an example of adaptation in one place at a particular time is not always applicable to a future adaptation at a different place, the insight may be gained about how the adaptation process may unfold in the future.

Analogues of technological innovation in agriculture and their relevance to climate change and adaptation include deliberate translocation of crops across different agro-climatic zones, substitutions of new crops for old ones, and substitutions of technology induced by scarcity of resources. Examples of crop translocation include thermal expansion of hard-red winter wheat in the Great Plains region of the United States and Canadian Prairie (Rosenberg 1992), and the expansion of the northern boundary of winter wheat in China, aided by the introduction of freeze-resistant varieties from Russia and Canada (Lin 1997, Chen and Libai, 1997). According to National Research Council (NRC, 1999), adaptation of canola in Canadian agriculture in the 1950s and 60s also shows the rapidity with which farmers and their supporting institutions adapt to market signals.

In response to the reduction of a frost-free growing season, farmers in Australia have modified their choice of cultivars in wheat (Meinke et al., 2003). For example, in Central Queensland, the average frost period shrunk from approximately 80 days at the end of 19th century to about 17 days at the end of 20th century. In response to such a change in growing season length, wheat in this region is sown earlier than they were in the 1950s and 1960s, targeting flowering dates of early to mid August to maximize grain filling period. Likewise farmers in the semiarid tropics of Kenya and Ethiopia have been able to increase water use efficiency through a combination of water harvesting techniques and drip irrigation that have enabled them to diversify cropping systems and minimize risk from increasing drought spells and erratic rainfall pattern (Ngigi et al., 2000). These examples provide evidence of adaptive research specifically targeted to address location-specific climatic needs. Yet, notwithstanding this recognition, there is a dearth of research that explains the role of climate as a stimulus for innovation of appropriate technologies (Ruttan, 1996).

5. Agricultural adaptation and the hypothesis of induced innovation

For the purpose of this paper, the term agricultural adaptation, following Easterling et al. (2004), refers to the actions that farmers and their supporting institutions take to reduce impacts or take advantage of new opportunities that may arise as a result of climate and other ongoing changes. Nearly two decades of research show that without adaptation, climate change is generally problematic for agricultural production (e.g., Reilly et al., 1996; Gitay et al., 2001; Easterling et al., 2007). Adaptation, in general, may lessen future yield losses (Easterling et al., 2007) or may improve yield gain in regions where new opportunities brought about by changing climate (e.g. extended growing season) have been seized upon. An overall synthesis by the Assessment Report Four (AR4) of the Intergovernmental Panel on Climate Change (IPCC) that plotted yields of three cereal crops (rice, wheat, and maize) against degrees of average local warming and associated changes in CO_2 shows that moderate to medium change in local temperature (1-3^{0}C) could have a small positive effect in yield in mid-to high-latitude regions. In the low-latitude regions simulation studies showing even moderate temperature change (1-2^{0}C) would likely have a negative effect on crop yields (Easterling et al., 2007). Although vulnerability of agriculture, defined here as the extent to which change in climate may damage food production from climate change net of adaptation, is greater in developing countries as compared to the developed countries, in both cases adaptation clearly ameliorates losses in yield. This reasoning has made the case for adaptation even stronger with clear indications that adaptation can only be ignored to the detriment of food security.

Articulated by Hayami and Ruttan in the early 1970s, the hypothesis of induced innovation has earned wide recognition in the field of agricultural development. It refers to the process by which societies develop technologies that facilitate the substitution of relatively abundant factors of production for relatively scarce factors in the economy. The hypothesis posits that the development of new technologies in agriculture is a continuing process induced by differences in the relative scarcity of resources, and is signaled by change in resource endowments. Furthermore, this hypothesis has emerged as a basis for understanding potential future agricultural adaptation to climate variability and change. Study by Chhetri and Easterling (2010) show an evidence of climate-induced innovation supporting the assertion that if farmers and public institutions are engaged in "co-production" of agricultural technologies, they are able to respond to climatic challenges much more readily than would have been possible otherwise. It is reasonable, therefore, to argue that the process of technological change represents an essential element of agricultural adaptation to climate change. As shown in Figure 1, climate change may alter these climatic resources by changing growing season length and soil moisture regimes, and by adding heat stress to the plant. Such changes, following the hypothesis of induced innovation, will provide appropriate signals to farmers and public institutions to induce technologies suitable for the new environment. Eventually, farmers and their supporting institutions will develop technologies to avoid the deleterious effects of climate change in agriculture. Some of these new technologies, however, may have positive feedback to climate systems and other could have negative feedback. The strength of this simple framework lies in its ability to highlight the central role of climate as a motivator of technological innovation and ultimately as a source of adaptation. Within this conceptual framework, it is important to examine the role of spatial and temporal variability in climate as an incentive to the innovation of technologies.

The hypothesis of induced innovation gained prominence with the publication of Yujiro Hayami and Vernon Ruttan's book, *Agricultural Development: An International Agricultural Perspective in 1971 and 1985*. In their work Hayami and Ruttan (1985) provide a powerful insight into the process of technological innovation in agricultural development. An important revelation into their analysis is the decomposition of changes in output per unit of labor into change in output per unit of land. These two factors of production are seen as "relatively independent" and associated with two different paths - one typified by the experience of the United States where progress in mechanical technology facilitated the substitution of power for human labor, and the other typified by the Japanese experience, where the progress in biological technology increased the productivity of land (p. 171).

The different paths of technological change in agriculture for the U.S. and Japan were shaped by differences in relative scarcity of resource endowments (labor and land). Throughout the period of 1880-1980 Japanese farmers used more fertilizer per hectare than the U.S. farmers. The U.S. farmers, in turn, used more machinery per worker than the Japanese farmers. In Japan, where land is relatively scarce, it was progress in biological technologies that led to increased response of rice varieties for higher level of fertilizer application. But in the U.S., where labor is a relatively scarce factor, it was the process of mechanization, first with animal and later with farm machinery (e.g., combine harvester), which facilitated the expansion of agricultural production by increasing the area operated per worker. In both cases, a process of dynamic adjustment to change in resource endowment was achieved through the innovation of appropriate technology.

To date, the hypothesis of induced innovation has been used to explain the relationship between resource endowment and the development of new technologies. There is now a substantial body of literature documenting the process of induced technological change in both developed and developing countries, and particularly the development of alternative technological trajectories to facilitate the substitution of relatively abundant factors for relatively scarce ones (e.g. Thirtle and Ruttan, 1987; Islam and Taslim, 1996; Thirtle et al., 1998). Studies on land and labor productivities for a broad range of counties and major geographic regions are broadly consistent with the hypothesis of induced innovation.

While the hypothesis of induced innovation has been credited with responding appropriately to the factor of production in agriculture, the role of climate as a stimulant for technological innovation remains largely unexplored. Little is known about the manner in which technology has altered the relationship between climate and society and the roles that climate has played in development of the new innovations. The interaction between climate and technology depends upon whether new technology encourages capital to be a complement or a substitute for climate (Mendelsohn et al., 2001). For example, if the marginal productivity of technology is higher for farms in an ideal climate, technology and climate could complement each other. In this case, technology is targeted to the ideal climate, making farmers in this environment more productive than farmers in marginal locations. However, if marginal productivity of technology is equal or better in a relatively unfavorable climate compared to the favorable one, then technology is said to have substituted for climate. Following the assertion made by the hypothesis of induced innovation, this is a desired condition if agriculture is to adapt to changing climatic resources in the future.

Although future climate is uncertain, the critical issue is how the research establishment has perceived the scarcity of climatic resources and how has this perception allowed them to integrate their attempts to innovate technology needed for farmers to respond to the evolving climate. Thus, the perception of resource scarcity, in this case climate, enters the Research and Development (R&D) process through interaction between farmers and the public institutions responsible for generating technologies. In the context of climate, mechanical innovations include irrigation, conservation tillage, leveled terracing, and integrated drainage systems - all of which are essential to widen agricultural activities that existing climatic resources would have, otherwise, not permitted. Biological innovations, on the other hand, also have significant roles in enabling agriculture to adapt to a wider range of climatic conditions. For example, through investment in crop-improvement programs, societies can develop varieties that are resistant to pest and diseases or tolerant to heat and drought, all pivotal in ameliorating the impact of climate change. If the farmers and their supporting institutions have not acquired the capacity to innovate technologies specific to their resource endowments, they will find it difficult to adapt to new and emerging climate. At its core, agricultural adaptation is about the capacity of all stakeholders to shift strategies to develop food production systems that are robust enough to withstand change driven by climate and other stressors.

6. Conclusion

The IPCC AR4 (IPCC, 2007) concludes that the Earth is committed to at least as much warming over the next 90 years as was experienced in the 20th century, even if greenhouse gas emissions were immediately curtailed to 2000 levels. This means that agricultural adaptation to climate change is inevitable. Studies conducted over two decades have identified a range of potential agricultural adaptations options for farmers. Most of these are extensions of existing risk management activities such as alteration of planting times to avoid temperature stress during critical plant growth and development stage, targeted irrigation to reduce water stress, and breeding of drought resistant cultivars. Some of these have been assessed as having substantial potential to offset negative effects of climate change. However, there has been little evaluation of how effective these and other adaptations may be given the complex nature of farm decision-making, the likely diversity of responses within and between regions, the possible interactions between different adaptation options, and the economic, institutional and cultural barriers to change.

According to the hypothesis of induced innovation, advances in knowledge can permit the substitution of more abundant resources for increasingly scarce resources to reduce the constraints for agricultural production. For example, innovation of early maturing cultivars has the greater potential of escaping the effects of drought that would be increasingly important to address the limitation of water scarcity due to a change in rainfall pattern. This may be done through the development of crop varieties that can perform even better with short supplies of water; through employment of water conserving techniques, such as conservation tillage farming; and/or through the development of efficient irrigation techniques, such as drip irrigation that maximizes the use of available water. In response to inadequate and or unreliable precipitation, it is reasonable to assume the same responses from farmers and their supporting institutions (Glantz and Ausubel, 1988). Characteristics of ongoing mechanisms used by farmers and their supporting institutions to manage their agricultural systems do provide information about the processes by which future adaptation may take place in the face of climate change.

If the systems of food production is thriving, either because farmers and their supporting institutions are modifying their strategies in ways that respond to emerging changes or because the underlying systems on which their livelihoods are based are sufficiently flexible to absorb the impact of climate changes, then both are considered to be robust. Yet, the current trend of weaning resources away from agricultural and climate research, especially in developing countries, endangers the vital support provided by public institutions for farmers to adapt to climate change. Therefore, agricultural adaptation to future climate is contingent upon continued investment in agriculture, as well as active engagement of public institutions responsible for developing and disseminating appropriate technologies for farmers operating in specific climatic regions. Successful adaptation involves a dynamic process of adjustment to resource endowment created by a new and changing climate. I argue that if research establishment and farmers have made appropriate responses to improve their capacity to respond to climatic constraints then they are generally better prepared to adapt to changing climate.

7. References

Adger, W.N., S. Agrawala, M.M.Q. Mirza, C. Conde, K. O'Brien, J. Pulhin, R. Pulwarty, B. Smit and K. Takahashi, 2007: Assessment of adaptation practices, options, constraints and capacity. *Climate Change 2007: Impacts, Adaptation and Vulnerability. Contribution of Working Group II to the Fourth Assessment Report of the Intergovernmental Panel on Climate Change,* M.L. Parry, O.F. Canziani, J.P. Palutikof, P.J. van der Linden and C.E. Hanson, Eds., Cambridge University Press, Cambridge, UK, 717-743.

Adger, W. N., Eakin, H. & Winkels, A. (2009). Nested and teleconnected vulnerabilities to environmental change. *Frontiers in Ecology and the Environment*, 7:150-157. Brush, S.B. and B.L. Turner, 1987. The nature of farming systems and views of their change. In: Turner B.L. and S.B. Brush (Eds.), *Comparative Farming*. The Guilford Press, New York, pp 11-48.

Brush, S.B. and B.L. Turner, 1987. The nature of farming systems and views of their change. In: Turner B.L. and S.B. Brush (Eds.), *Comparative Farming*. The Guilford Press, New York, pp 11-48.

Chen, H. and H. Libai, 1997: Investigating about varieties filtering in winter wheat northward shifting. *Journal of Chenyang Agricultural University*, 28:175-179.

Chhetri, N. and W.E. Easterling. 2010. Adapting to climate change: retrospective analysis of climate technology interaction in rice based farming systems of Nepal. *Annals of the Association of American Geographers* 100(5):1-20. DOI: 10.1080/00045608.2010.518035.

Cisse, N., Ndiaye, M., Thiaw, S., Hall, A.E., 1997. Registration of 'Mouride' cowpea. *Crop Science*, 35:1215-1216.

Easterling, W.E., 1996. Adapting North American agriculture to climate change in review, *Agricultural and Forest Meteorology*, 80:1-53.

Easterling, W.E., H.H. Brian, and J.B. Smith, 2004. *Coping with global climate change: the role of adaptation in the United States*. Pew Center on Global Climate Change, pp. 40.

Easterling, W.E., P.K. Aggarwal, P. Batima, K.M. Brander, L. Erda, S.M. Howden, A. Kirilenko, J. Morton, J.-F. Soussana, J. Schmidhuber and F.N. Tubiello. 2007. Food, fibre and forest products. *Climate Change 2007: Impacts, Adaptation and Vulnerability. Contribution of Working Group II to the Fourth Assessment Report of the Intergovernmental Panel on Climate Change*, M.L. Parry, O.F. Canziani, J.P. Palutikof, P.J. van der Linden and C.E. Hanson, Eds., Cambridge University Press, Cambridge, UK, 273-313.

Elawad, H.O.A. and A.E. Hall, 2002. Registration of 'Ein El Gazal' cowpea. *Crop Science*, 42:1745-1746.

Evenson, R.E. and D. Gollin, 2000. *The Green Revolution: End of Century Perspective*. Paper Presented for the Standing Project on Impact Assessment of the Technical Advisory Committee of the Consultative Group on International Agricultural Research (CGIAR). New Haven, Connecticut, U.S. Economic Growth Center, Yale University.

Fischer, G., Shah, M., Tubiello, F., and van Velthuizen, H. 2005. Socio-economic and climate change impacts on agriculture:An integrated assessment, 1990-2080. *Philosophical Transactions of Royal Society* B, 360:2067-83.

Gitay, H., Brown, S., Easterling, W. and Jallow, B. 2001. Ecosystems and their Goods and Services. In *Climate Change 2001: Impacts, Adaptation, and Vulnerability*, Eds. McCarthy, J. pp. 235–342. Contribution of Working Group II to the Third Assessment Report of the Intergovernmental Panel on Climate Change, Cambridge University Press.

Glantz, M.H. and J. H. Ausubel, 1988. Impact Assessment by Analogy: Comparing the Impacts of the Ogallala Aquifer Depletion and CO_2-Induced Climate Change. In *Societal Responses to Regional Climatic Change: Forecasting by Analogy*, Eds. Glantz, M.H, pp.113-142.Boulder, Westview Press,

Hall, A.E., 2004. Breeding for adaptation to drought and heat in cowpea. *European Journal of Agronomy*. Available online @ www.sciencedirect.com.

Hayami Y. and V W. Ruttan, 1971. *Agricuitural development: an international perspective*. The John Hopkins University Press, Baltimore, pp. 367.

Hayami, Y. and Ruttan V. W., 1985. *Agricultural development: An international perspective*. The John Hopkins University Press, Baltimore, pp. 506.

Heyd, T. and N. Brooks, 2009. Exploring cultural dimensions of adaptation to climate change. In N. Adger, I. Lorenzoni, and L. O'Brien (Ed.), *Adapting to Climate Change: Thresholds, Values, Governance* (pp 269-282). Cambridge, UK: Cambridge University Press.

Islam, T. and M. A. Taslim, 1996. Demographic Pressure, Technical Innovation and Welfare: The case of the agriculture of Bangladesh. *The Journal of Development Studies*, 32, 734-770.

Jodha, N.S., 1978. Effectiveness of farmers' adjustment to risk. *Economic and Political Weekly*, 13:A38-A48.

Lin, E., 1997. *Modeling Chinese agricultural impacts under global climate change*. Chinese Agricultural Science and Technology Press, Beijing, China, pp: 61-68

Liverman, D.M., 1990. Drought impacts in Mexico: climate, agriculture, technology, and land tenure in Sonora and Puebla. *Annals of the Association of American Geographers*, 80:49-72.Meinke, H., W. Wright, P. Hayman, and D. Stephens 2003. Managing cropping systems in variable climates. In *Principles of field crop production (IV Edition)*, Eds. Pratley J. pp. 26-77. Oxford University Press, Melbourne, Australia.

Meinke, H., W. Wright, P. Hayman, and D. Stephens 2003. Managing cropping systems in variable climates. In *Principles of field crop production (IV Edition)*, Eds. Pratley J. pp. 26-77. Oxford University Press, Melbourne, Australia.

Mendelsohn, R., A. Dinar, A. Singh, 2001. The effect of development on the climate sensitivity of agriculture. *Environment and Development Economics*, 6:85-101.

National Research Council (NRS), 1999. *Human dimensions of global environmental change: Research pathways for the next decade*. National Academy Press, Washington D.C. pp. 83.

Ngigi, S.N., J.N. Thome, D.W. Waweru, and H.G. Blank, 2000. *Technical evaluation of low-head drip irrigation technologies in Kenya*. Research report, Nairobi University and the International Water Management Institute (IWMI), Nairobi, pp. 21.

Pinstrup-Andersen, P., 1982. *Agricultural Research and Technology in Economic Development*. New York, Longman.

Reilly, J., 1996. Agriculture in a changing climate: Impacts and adaptations. *In* Houghton, J. T., L. G. Meiro Filho, B. A. Callander, N. Harris, A. Kattenberg, and K. Maskell, eds, *Climate Change 1995: The Science of Climate Change. Contribution of Working Group I to the Second Assessment of the Intergovermental Panel on Climate Change*. Cambridge University Press, Cambridge, pp. 584.

Rosenberg, N.J., 1992. Adaptation of agriculture to climate change. *Climatic Change*. 21:385-405.

Ruttan, V.W., 1996. Research to achieve sustainable growth in agricultural production into the 21st century. *Canadian Journal of Plant Pathology*. 18:123–132.

Smithers, J. and A. Blay-Palmer, 2001. Technology innovation as a strategy for climate adaptation in agriculture. *Applied Geography*, 21:175-197.Thirtle, C., R. Townsend, van Zyl, J. 1998. Testing the induced innovation hypothesis: an error correction model of South African agriculture. *Agricultural Economics*, 19, 145-157.

Thirtle, C. G. and V. W. Ruttan, 1987. The role of demand and supply in the generation and diffusion of technical change. Harwood Academic Publication, 173 pp.

Thirtle, C., R. Townsend, van Zyl, J. 1998. Testing the induced innovation hypothesis: an error correction model of South African agriculture. *Agricultural Economics,* 19, 145-157.